An Introduction to
AGRICULTURAL SYSTEMS

Second Edition

An Introduction to
AGRICULTURAL SYSTEMS

SECOND EDITION

C. R. W. SPEDDING

*Professor of Agricultural Systems and Director of Centre for
Agricultural Strategy, University of Reading, UK*

ELSEVIER APPLIED SCIENCE
LONDON and NEW YORK

ELSEVIER APPLIED SCIENCE PUBLISHERS LTD
Crown House, Linton Road, Barking, Essex IG11 8JU, England

Sole Distributor in the USA and Canada
ELSEVIER SCIENCE PUBLISHING CO., INC.
52 Vanderbilt Avenue, New York, NY 10017, USA

WITH 64 TABLES AND 52 ILLUSTRATIONS

© ELSEVIER APPLIED SCIENCE PUBLISHERS LTD 1988

First edition 1979
Second edition 1988

British Library Cataloguing in Publication Data

Spedding, C. R. W. (Colin Raymond William)
An introduction to agricultural systems.—
2nd ed.
1. Agricultural systems
I. Title
630

Library of Congress Cataloguing-in-Publication Data

Spedding, C. R. W.
An introduction to agricultural systems.

Bibliography: p.
Includes index.
1. Agricultural systems. I. Title.
S494.5.S95S64 1988   630   88-385

ISBN 1-85166-191-3

Printed in Great Britain by Galliard (Printers) Ltd, Great Yarmouth

Preface to the Second Edition

It would have been very easy to expand on all the sections of the first edition but I decided to try to retain the relatively short, introductory nature of the book.

Some new material has been added, particularly where it has been possible to update data, and there has been some change of emphasis in places, in order to reflect changing world conditions.

The book retains its original purpose, however, of introducing systems thinking as applied to agriculture.

I am grateful to Angela Hoxey for help in preparing this edition, especially in relation to the preparation of tables and figures.

C. R. W. SPEDDING

Preface to the First Edition

The agricultural systems of the world represent a very large subject. Their study involves a great deal of fairly detailed knowledge, as well as a grasp of the structures and functions of the systems themselves. This book has been written as an introduction to such a study and it concentrates on an overall view, rather than on the detail, partly because of the need to relate the latter to some larger picture in order to appreciate the relevance and significance of the detail.

This problem—of seeing the relevance of component studies and the significance of physical, biological and economic detail, and indeed principles—is encountered by many agricultural students right at the beginning of their university careers.

A systems approach to the teaching of agriculture aims to help by providing a framework that can be clothed with detail as time and interest allow. The initial framework has to be simple and usable but it is also important to think clearly about the reasons for studying the subject and the ways in which this can be done. By considering all these matters, I hope this book will serve as a useful introductory text in the teaching of agriculture as a whole.

C. R. W. SPEDDING

Contents

Acknowledgements

It is a pleasure to acknowledge the help of many colleagues, but especially that of Miss A. M. Hoxey, Dr J. M. Walsingham, Dr G. P. Harris, Mr A. K. Giles, Mr J. M. Stansfield and Mr P. M. Harris, in the preparation of this book.

I have to thank Miss Hoxey for drawing Figs 1.4, 2.3, 2.11, 2.13–2.16, 3.1, 5.4, 7.6, 11.3 and 12.2 and Dr G. R. Spedding for drawing all the remaining figures.

I am grateful to Mr L. Norman, Mr R. B. Coote and Longmans for permission to reproduce Fig. 10.1 and to Dr E. J. T. Collins and Dr J. G. W. Jones for helpful comments on the manuscript.

I am most grateful to Brookwick, Ward and Co. Ltd (London) for permission to reproduce two photographs of their Hauptner models as Figs. 2.6 and 3.1, and to Miss L. M. Spedding for preparing the index.

I am particularly grateful to my secretary, Mrs Valerie Craig, for her accurate typing and for much other help during the whole period in which this book was being prepared, from inception to publication.

1

The Purposes of Agriculture

The first point to clarify is: 'What is agriculture?' Of course, there is general agreement about the sorts of things, people, plants and animals that can be called agricultural, but this is not good enough if we are seriously interested in topics such as the role of science in agriculture, the role and importance of agriculture in the world and how agricultural efficiency can be improved.

Not many attempts have been made to be more precise and it is quite difficult to arrive at a definition that is both useful and specific. Incidentally, there are many useful 'umbrella' terms that can and do cover a multitude of different things and their usefulness may be reduced if they are too rigidly defined. 'Competition', 'growth' and 'animal' are examples. The only satisfactory definitions are very broad and do not really exclude everything that does not strictly belong; stricter definitions, on the other hand, exclude things that one wishes to include. The best solution in these cases is to attach strict definitions to classified or qualified groups *within* the big category. Thus, 'growth in length' is more specific, as is 'competition for light': 'cold-blooded animal' or, for example, 'unicellular animal' both become more specific by the use of an adjective.

However, it is a worthwhile and challenging occupation to try and define anything we wish to discuss, provided that we remember to make the definition a useful one. By 'useful' is meant that it enables us to distinguish between the thing defined and other things; in this case, to distinguish 'agricultural' from 'non-agricultural' things.

Agriculture itself is clearly an *activity* and, furthermore, it is an activity of Man. If you disagree with this, you have to include the fungus growing carried out by certain ant species (about 100 species of Myrmecine ants and some species of termite (*Termes* spp.)) and argue that it is the activity that matters and not who is doing it. Well, why not? On balance it seems to me

1

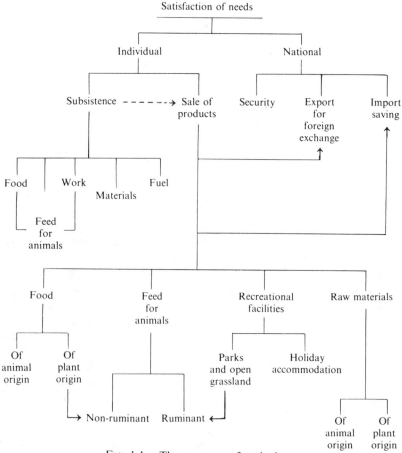

FIG. 1.1.   The purposes of agriculture.

slightly more useful to limit it to activities of Man—in which case the ants are engaged in an activity similar to agriculture—but it could easily be argued that activities are independent of the name and nature of those who carry them out.

What, then, is the nature of the activity? There are two main ways of establishing this: the first is to consider *why* the activity is carried out and the second is to consider what is actually (and characteristically) done.

The first question is the more difficult. Agriculture is carried out for a great variety of purposes (illustrated in Fig. 1.1) but most, if not all of them, are associated with producing *products*. The difficulty comes when we try to define these products. They include animal and plant materials, dead and alive, some for food, some for clothing and furnishings, some for fuel (e.g. cattle dung) and some for shelter. The animals used may not be very different from those that are hunted and the plants may be similar to some wild species. Fruit trees, rubber trees and nut trees are usually included but forest trees are not: the logic of this is difficult to grasp but forestry is generally regarded as a separate subject. The simplest solution is either to exclude timber production from the definition of agriculture or to regard forestry as a separate branch of agriculture. It is still worth asking, of course, why it should be separate and whether the traditional usage is necessarily right. (You should have a view about this, either now or by the end of this chapter.)

The main products may be described as *food, fibre* or *raw materials for industrial use* and they are produced by the keeping of animals and the cultivation of crops.

The second question seems fairly straightforward, therefore, and can be answered along these lines. The activities that are characteristically agricultural are the keeping of animals and the cultivation of crops. Many people would prefer to regard these as primarily land-using activities and would rather regard as non-agricultural those animal and crop production systems that appear to be divorced from the land. Thus, housed poultry and hydroponics (the growing of plants in soil-less solutions) seem less agricultural than grazing cattle and growing fields of wheat. This does not seem very logical, however, and would exclude winter housing of stock and watercress production, for example.

More important, in arriving at a definition, is to emphasise the fact that if the animals and plants are not under some degree of control, then the situation could be similar to hunting and food collection. The degree of control required in agriculture is very variable but it usually applies: (a) to the physical location of the organism; (b) to its nutrition; (c) to its reproduction and (d) to the method and pattern of harvesting, to a greater or lesser extent.

Some animals are easier to control than others and this may account for the relatively few species used in agriculture, although the same argument can hardly be used to explain the situation for plants (see Table 1.1). The main agricultural plants and animals are given in Tables 1.2 and 1.3. Most are terrestrial and the reasons for this may include problems of control in water, including the problem of controlling nutrition. It is quite difficult, for example, to control plant nutrition in the sea, because of the huge volume of water that dilutes everything added to it.

Most of the animals are warm blooded and of a fairly large size. The reasons for this presumably relate to ease of domestication and the relative usefulness of different animals to Man at the time when domestication was taking place. It seems likely that the earliest attempts at domestication would have been directed at those animals that were hunted and those plants that were gathered for food. It also seems likely that

TABLE 1.1

NUMBER OF SPECIES USED IN AGRICULTURE

(The total number of plant and animal species is vast: only higher plants and warm-blooded animals are included here)

| | Crop plants[a] | Agricultural animals[b] | |
		Mammals	Birds
Number of species used	1 000–2 000 (perhaps 10 000–20 000 are *cultivated*)	20–30	5–10
% of those available (i.e. higher plants and warm-blooded animals)	c. 0·4	0·5–0·75	0·05–0·1
Number of species of major importance	100–200	c. 10	c. 3
Number that provide most of the world's food	15	5	2

[a] Based on Janick *et al.* (1969).
[b] Based on Zuener (1963), Morris (1965), Farmer and King (1971).

domestication would have proceeded most rapidly where reproduction occurred readily. This would apply to animals that mated and reproduced in some sort of captivity or under some degree of control and to plants in which the reproductive organs were eaten and stored. In the latter case, with seeds and tubers, for example, another crop could arise by accident but in such a way that the opportunity offered would be obvious.

It may seem surprising that insects were not used to a greater extent, especially amongst peoples known to eat them, since they are relatively easy to control, feed, breed and harvest. In fact, only two insects can really be said to be 'farmed', the honey-bee (*Apis mellifera*) and the silk moth (*Bombyx mori*), neither of which are eaten.

The activity represented by agriculture is thus a purposeful one, with the main aim of producing products, and involves the controlled use of characteristic animals and plants. Since agriculture must clearly aim to produce more (money, energy, goods) than it uses in the production

TABLE 1.2
THE MAIN CROP PLANTS

Crop category	Examples
Cereals	Maize, rice, wheat, oats, barley, millet
Pulses	Bean, pea, peanut, soyabean, cowpea
Forage crops	Grass, clover, lucerne (alfalfa)
Roots and tubers	Potato, cassava, sweet potato, turnips
Leafy crops	Cabbage, spinach
Fruits	Orange, lemon, lime, olive, apple, strawberry
Oil crops	Palm, peanut, olive, cottonseed, linseed, sunflower
Nuts	Almond, filbert, pecans
Sugar crops	Sugar cane, sugar beet
Beverage, spices, etc.	Coffee, tea, cocoa, grape, perfumes, peppers
Fibre crops	Flax, jute, hemp, sisal, cotton
Fuel crops	Hardwoods, softwoods

process, it is always concerned with the sensible use of resources. It is thus essentially an *economic* activity, even where output is not expressed in monetary terms.

A definition embracing these features might then be phrased as follows:

'Agriculture is an activity (of Man), carried out primarily to produce food and fibre (and fuel, as well as many other materials) by the deliberate and controlled use of (mainly terrestrial) plants and animals.'

TABLE 1.3
THE MAIN AGRICULTURAL ANIMALS
(See Table 12.1 for more detail)

Mammals	Horse
	Ass
	Mule
	Camel
	Cattle
	Buffalo
	Sheep
	Goat
	Pig
Birds	Domestic fowl
	Duck
	Goose
	Turkey
Cold-blooded vertebrates	Fish
Invertebrates	Bee
	Silk moth

TABLE 1.4
THE MAIN CROP PRODUCTS

Food (for humans)	Cereals
	Starchy roots
	Sugar
	Seeds especially pulses
	Nuts
	Oils
	Vegetables
	Fruits
	Beverages
	Flavourings
Medicines	Quinine, opium, cocaine
Fumitories	Tobacco
Masticatories	Betel nuts
Feed (for animals)	Fresh green feed
	(grass and forages)
	Conserved feeds
	(hay, silage)
	Roots
	By-products
	(bagasse, straw, beet pulp)
	Concentrates
	(cereals, oil seed cake)
Materials	
Construction	Timber
Paper	
Textiles	Cotton, hemp
Rubber	
Household goods	Cork, woven utensils
Fertilisers	Green crops
	Crop wastes
Fuel	Wood, charcoal, alcohol, methane
Industrial oils	Linseed, cottonseed, corn oil
Essential oils	Perfumes, camphor
Gums	Gum arabic, gum tragacanth
Resins	Lacquer, turpentine
Dyes	Logwood, indigo, woad
Tannins	Hemlock, oak, mangrove, wattle
Insecticides	Roterone, pyrethrum

This would exclude gardening and landscaping unless products could be described for them (such as money?) but forestry, fish farming and a number of industrial processes would be included.

The word 'primarily' implies that there are other important products and this is indeed so: the main crop and animal products are listed in Tables 1.4

and 1.5. Of course, not all crop—or, indeed, animal—products are directly used by Man; many are grown to feed to animals, so the 'production of food' has to include food for animals as well. Alternatively, feed† for animals can be regarded as an intermediate product, or feed production as a process that takes place *within* an agricultural system.

TABLE 1.5
THE MAIN ANIMAL PRODUCTS

Food	Meat
	Milk
	Eggs
	Fish
	Honey
	Blood
Fibre	Wool
	Hair
	Fur
	Silk
Skins	Leather
Fertiliser	Faeces
	Bone
	Feathers
	Horn
Work	Transport
	Traction

Definitions are never as permanent as they sound and there is no reason to retain one (including that proposed here) if a more useful version can be found. Furthermore, we must always be careful not to accept too readily what seems obvious.

This brings us to a common difficulty of definitions. Although a cow is obviously an agricultural animal, it is really only so when it is being used agriculturally (i.e. for food production, for example). Even accepting that animals do have other agricultural functions and roles than just food production, a pet cow is clearly *not* an agricultural animal. Similarly, there is no difficulty at all in accepting that the horse may or may not be an agricultural animal, just as is the case for land, people, plants and water.

These differences and distinctions are of great importance where subsidies, laws or regulations, for example, may only apply to activities regarded as 'agricultural'. Furthermore, these distinctions become

† The word 'feed' is sometimes used to distinguish 'feed' for animals from 'food' for humans.

increasingly difficult to make in those countries (such as the UK) where farming is becoming integrated with non-farming enterprises. Recreational activities on farms may use horses, fish production may be undertaken to provide sport for anglers, rare breeds may be kept because visitors like to look at them and crops may be grown as cover for game birds. For these reasons, the range of crops and animals that are regarded as 'agricultural' may increase.

So particular animals (species, breeds or individuals) and particular plants (species, varieties or individuals) may serve as illustrations of agricultural organisms but whether they are themselves agricultural or not depends entirely on whether they are embedded in agricultural (processes or) systems or not: and processes are only agricultural when embedded in such systems. This is one reason why the next chapter deals with agricultural systems and the problems of what constitutes a system and how we know when we are only looking at a part of one.

The agricultural activity itself, of course, may only be a part of, for example, national life. Similarly, the activity that we have defined agriculture to be can also be studied and taught, as well as practised. The subject studied would also be called agriculture—the study of it does not have a special word, equivalent to zoology for the study of animals or botany for the study of plants. The practice of agriculture is generally called farming and occurs most usually in a land-use context.

The term 'agricultural science' is sometimes used to describe the scientific study of agriculture but this can be rather misleading because agriculture embraces more than the contributing sciences and cannot, therefore, necessarily be studied scientifically in all its aspects. (A similar difficulty applies, for example, to 'social science'). 'Agricultural science', on the other hand, can be used to group those sciences (or parts of them) that are most relevant to agriculture or underpin its processes and operations.

Since agriculture is an activity undertaken by, and involving, people, however, and is carried out for productive (and profitable) purposes, it must necessarily include aspects of economics and management, as well as biology. The multi-disciplinary nature of the subject is, to many, one of its major attractions, partly because it is not confined by arbitrary boundaries and partly because it deals with real life situations, characterised, as they so often are, by being mixtures of different disciplines.

This is true even of deceptively simple examples. A cow, for example, is an animal but it lives on grass, the growth of which it affects by biting, treading, lying, dunging and urinating on it. Furthermore, it is only able to digest it because it has, in one large compartment of its stomach (the rumen), vast

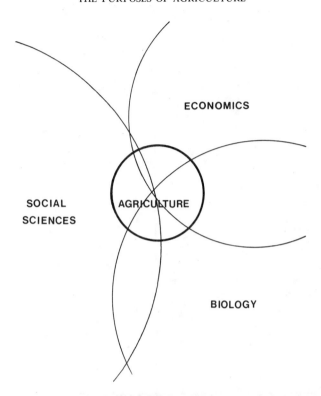

FIG. 1.2. Agriculture and the overlap of disciplines: for simplicity only three major disciplines are shown (overlaps with engineering, veterinary medicine and human medicine also occur).

numbers of minute plants (bacteria) that are capable of digesting cellulose. The cow is thus a ruminant, and used as such agriculturally, and the definition of a ruminant includes the presence of plant populations in the stomach.

How, then, could a cow be completely studied by a zoologist or a botanist alone? In farming, however, the cow also costs money to purchase, house, feed and milk, and if its products did not realise more money than these production costs, farming would cease. So it would if people found the conditions of milking intolerable or cows died of disease or no-one wanted milk anyway. Many subjects are, *of necessity*, involved in, and contribute to an understanding of agriculture. The main overlaps between contributing subjects are shown in Fig. 1.2.

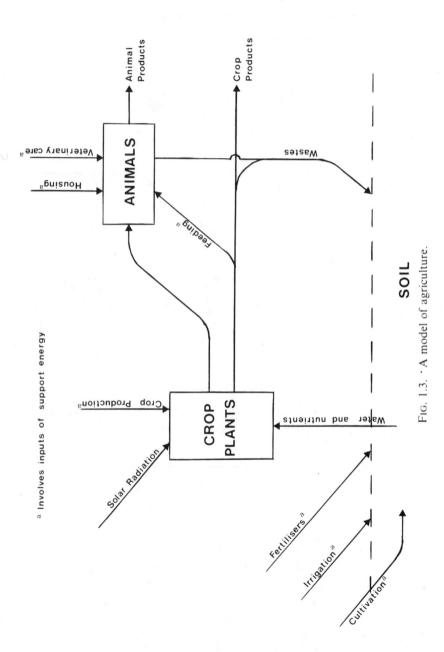

Fig. 1.3. · A model of agriculture.

Agriculture as a subject, therefore, is concerned with an activity of fundamental importance to all communities and consists of a purposeful blend of science and non-science.

AGRICULTURAL MODELS

Before going on to discuss agricultural systems (the operational units of agriculture) and the models needed to describe them, it is reasonable to ask whether a picture of 'agriculture' or the 'agricultural activity' can be produced. After all, the essential agricultural activity should be describable, otherwise we cannot make or discuss propositions about it.

The purpose of such a picture is clear: it is to amplify the definition, so that we have a better, more comprehensive, more detailed and more precise description of agriculture.

Figure 1.3 is an attempt at this or, rather, at trying to convey what such a picture might or should look like. Pictures intended to illustrate or demonstrate have to be fairly simple, in order to be clear: working drawings, on the other hand, require immensely more detail and are not easy to reproduce in a visually satisfactory form. If you wanted to use such a picture for a working purpose, therefore, such as deciding what further research was needed or how agriculture could be improved, you might devise much more appropriate figures. However, many of these would only have to deal with a part of agriculture and our attempt here is to generalise about agriculture as a whole. One way of doing this has been to consider the matter historically and to trace the way in which agriculture has developed to meet human needs.

THE ORIGINS OF AGRICULTURE

It is possible to deduce some of the early developments in agriculture from archaeological remains. For our present purposes, however, it is only necessary to note that organised agriculture must have been preceded by food gathering and hunting, both of which still persist to a greater or lesser extent. Such methods of obtaining food do not allow many people to live per unit area of land (Table 1.6) and the need to increase food production clearly stems from the increased needs associated with settlement.

The cultivation of crops and the keeping of animals require consider-able control over the reproduction of both, over their pests, parasites

TABLE 1.6

NUMBER OF PEOPLE SUPPORTED PER SQUARE
MILE (P)
(Based on Harrison *et al.*, 1964)

	P
Food gatherers	Up to 2
Higher hunters and fishers	Up to 20
Simple cultivators	$\frac{1}{2}$ up to 50
Pastoralists and nomads	10 to 100
Advanced cultivators	10 to 150
Iron Age (Britain)	10
Middle Ages (Britain)	50

and predators, and over nutrient supplies. Many other developments must also have occurred. Various ways of using power (from wind, water and animals) would not only solve problems of cultivation, harvesting and transport, but would pose new problems, such as the invention and manufacture of equipment, harnesses and vehicles.

It does not require a great deal of thought to see how closely interwoven must have been the agricultural and non-agricultural aspects of human life during these early periods.

Gradually, however, as productivity per unit of land and per man rose, so only a small proportion of the population became essential for food production. This trend confers immense benefits but, in the past, has rested largely on the use of relatively cheap oil to make and power things that make it possible for one man to do a great deal more than he could manage without such aids. The consequences of this dependence on oil will be further discussed in Chapter 4: the implications for future development are considerable.

FUTURE DEVELOPMENTS

Existing techniques and methods in many parts of the world can be greatly improved. This is not necessarily a matter of more and more expensive or complicated equipment; sometimes it is simply a matter of more intelligent design. A good example of this can be found in the hoe used for centuries in Nigeria (and now being replaced). This has traditionally involved a very short handle and a very acute angle between the handle and the blade, requiring the operator to work bent double and resulting in a high incidence of back trouble.

Indeed, one of the features of the future must be *not* to assume that improvement requires complexity and great cost.

The areas of potential improvement are chiefly the organisms used, the nature and level of inputs, better planning, use and storage of outputs and cheaper sources of energy. Animal and plant breeding and selection are aimed at improving the organisms on which agriculture is based. Improvement here generally has to mean in directions that lead to greater biological (Chapter 4) or economic (Chapter 5) efficiency. Of course, it

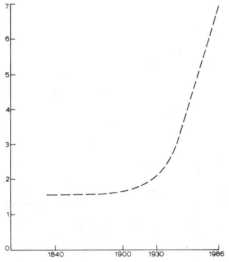

FIG. 1.4. Estimated increase in the UK wheat yield over the last 140 years. This has been associated with increasing use of artificial fertilisers over the last 50 years but many other changes, in the wheat varieties available and in farming methods, have also occurred, and it is not possible to dissociate them completely (x-axis, year; y-axis, tonnes of wheat/ha). (After Tatchell, 1976.)

cannot be assumed that we have necessarily chosen to use the best or only species that could be used and, in fact, we use very few of those that exist.

In the past, great increases in yield have resulted from large increases in the level of inputs such as fertilisers to crops (see Fig. 1.4). Some of these (especially fertilisers) are energetically costly to produce and greater dependence on the biological fixation of nitrogen may be expected to be sought in future. The problems are how to do this without too great a reduction in output per unit of land.

TABLE 1.7
ESTIMATED LOSSES IN WORLD AGRICULTURAL PRODUCTION
(After Duckham *et al.*, 1976)

	Energy	
	Per unit of cultivable land ($GJ/ha/y$)	*Per head of total population* (MJ/day)
Photosynthate actually formed in food production and estimated as 'recoverable'	25·8	59·2
Photosynthate recovered as potential food products	9·1	20·9
Food products entering households	4·1	9·5
Food products actually eaten	3·7	8·6

One of the most striking features of present day agriculture is the enormous discrepancy between what is produced initially and what is finally consumed. Losses occur on an enormous scale (see Table 1.7) and a major objective in the future must be to reduce them. They occur during crop growth—due to weeds, pests and disease—and during harvesting and storage: in animal production, similarly, pests, parasites, disease and inadequate nutrition result in very large losses, directly and of potential. In both cases, there are further losses in processing, preparation and cooking.

Cheaper sources of energy will need to be sought because agriculture does require sources of inputs and power, and oil is a costly and non-renewable resource. However, the energy available in solar radiation is vast and that in wind, water and agricultural waste materials very considerable.

Future developments are most likely, then, in these general directions, involving less waste, greater recycling and energy from renewable sources but not at the cost of too great a lowering of food output per unit of land.

2

A Systems Approach to Agriculture

The operational units of agriculture may be described as agricultural systems, including all the variations in size and complexity of unit that are called enterprises, farms, plantations, regional and national agricultures. It is useful to have a term of this kind that does not imply a particular size or level of organisation. After all, the essential activity does not imply anything of this kind and it is desirable to be able to discuss operational units in general. However, there is much more to the deliberate use of the word 'systems' in this context, because the 'systems approach' is now recognised as a distinctive way of looking at things. This is based firmly on the concept of a system and it is important to be clear about its meaning.

DEFINITION OF A SYSTEM

So many different things can legitimately be regarded as systems (a bicycle, a car, a cow, one's own body, a farm, a sewage works) that it is tempting to conclude that anything can be a system. This, however, is not so and, if it were so, the concept would be useless. If it were not possible to distinguish 'systems' from 'non-systems' the concept could not be used and if the distinction did not involve important properties it would not be worth making. It is the properties of systems that chiefly matter and they may be summarised in the phrase 'behaviour as a whole in response to stimuli to any part', Thus, a collection of unrelated items does not constitute a system. A bag of marbles is not a system: if a marble is added or subtracted, a bag of marbles remains and may be completely unaffected by the change. The marbles only behave as a whole if the whole bag is influenced, for example by dropping it, but if it bursts the constituent parts will go their own ways.

Of course, any collection can be transformed into a system by building into it such relationships between the components as are required to give it characteristic system properties. A simple, if rather trivial, example would be nailing together separate pieces of wood: once joined they would behave as a whole because of their physical connections.

However, not all sets of components that are joined together constitute systems. They may be joined up to other units and be incapable of behaving independently and thus be incapable of responding to stimuli at all. So a bicycle is a system but, if it was joined up to another one or to a sidecar, it would then be necessary to regard the new combination as a system and the original bicycle either as a sub-system (but this ought then to mean something special) or simply as part of a system.

An animal is a good example of a living system. It has an obvious structure, it behaves as a whole in response to major stimuli and it is relatively easy to see where it begins and ends (i.e. to distinguish the boundaries between the system and its non-system surroundings). This last point is very important. It is of only limited value to be able to identify a system if it is not possible to say where it ends. In the case of an animal, such as a hen, the obvious boundary would be just outside the external layer of

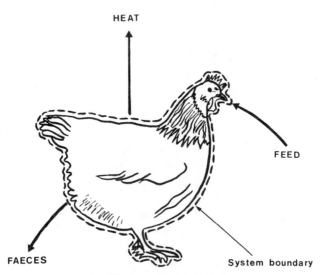

FIG. 2.1.   The hen as a biological system.

feathers, including the thin layer of air that forms a kind of 'private' micro-climate around the hen. Figure 2.1 illustrates this in an outline diagram, showing the major inputs (food and water) and the major outputs (excreta and heat). Now, in most situations these outputs do not immediately, directly or automatically influence the hen producing them. The atmosphere is so enormous that the heat output of the hen does not appreciably influence it. But suppose that the hen was confined in a small, sealed box. This immediately reminds us of the limitations of our model: for example, oxygen was not mentioned and will clearly run out, with disastrous consequences. However, let us confine ourselves to the heat output in order to illustrate the main point. The atmosphere in the box is soon heated up by the hen and immediately and directly affects the hen and the rate at which it produces heat. This is called a 'feedback' mechanism (or loop) and, if we ignore it, we are going to be misled about how our system (the hen) will respond to any stimulus. The hen in the box is no longer a

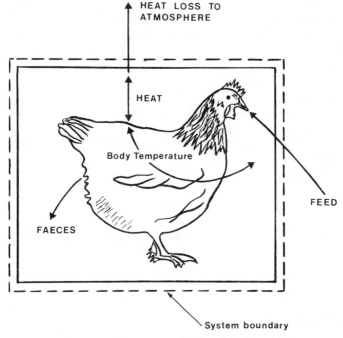

FIG. 2.2. The 'hen-in-a-box' system. The heat output of the hen now influences its micro-climate and thus its body temperature; this, in turn, may reduce feed intake and faecal output and lower the heat output of the bird.

sufficiently independent organism to be regarded as a system: it is necessary to consider the 'hen + box' and place the system boundary around the outside of the box (see Fig. 2.2). This correct positioning of the boundary is a way of defining the content of a system and, if it is not done, much of the value of a systems approach is lost.

Imagine that we studied different parts of the body (arms, legs, etc.) separately and in isolation. We would clearly never understand how the whole body worked, but we would not even understand how each part functioned *when joined up to the body*: and it is precisely in these latter circumstances that we particularly wish to understand the function of parts. In fact, they cannot function separately but, in any of the senses in which they might, it would be quite irrelevant to their role as parts of the body system.

Thus it is important to know when we are dealing with a whole system and when we are dealing with a part. Since it is not possible to study whole systems all the time, however, it is useful to be able to identify 'sub-systems', i.e. parts of systems that are worth considering separately because their separate function *is* relevant to their role in the whole system. (We will return later to this question of sub-systems.)

It is this importance of being able to tell when we are dealing with a system that makes the definition vital. There have been a great many attempts to define a system satisfactorily; the following contains the main elements discussed above:

> 'A system is a group of interacting components, operating together for a common purpose, capable of reacting as a whole to external stimuli: it is unaffected directly by its own outputs and has a specified boundary based on the inclusion of all significant feedbacks.'

A SYSTEMS APPROACH

This way of looking at the world and of tackling problems is founded on the idea that it is necessary to identify and describe the system that one wishes to understand, whether in order to improve, repair or copy it, or to compare it with others in order to choose one.

Its main proposition is obvious and is common sense but it is not always clear how to operate on this basis in practice. Certainly the approach does not come naturally or automatically and has to be cultivated. Fortunately, it is now backed up by techniques and methods that are very powerful, but this also means that they can be misused disastrously. So, as with all powerful tools, it is necessary to accept a learning period and to be very clear at the outset what it is we are trying to do and how we propose to do it.

A learner pilot does not jump into an aircraft and expect to fly immediately: even less does he dissipate his energies trying to flap the wings up and down because he has observed birds doing this. It is, in fact, quite an art to be able to use tools skilfully, for it is necessary to accept the limitations of the tool and to use it when and where it is appropriate. The danger is then that one may only do what the tool allows, even when new tasks become more important. The tool-user has to be aware of the developing needs for tools and the development of new tools by those who concentrate on such innovation. A systems approach embraces the disciplined use of existing tools, the development of new ones (which involves studies of methodology) and the definition of tasks for which tools are required.

Since a systems approach can be applied to many subjects, and not just to agriculture, its characteristics tend to include methods and techniques. But many of these also apply to other approaches, so the essence is, in fact, more of a philosophy (it is well named an 'approach', or a way of doing things and thinking about things).

The central idea that one must understand a system before one can influence it in a predictable manner, has embedded in it the importance of recognising a system, what is in it and what is not, and where its boundary lies. This is all included in the identification and description of a system and this represents a kind of model building.

MODELS

There are all kinds of models, but they are all representations of the real thing, simplified for some purpose: they include those features that are essential for the purpose and they leave out those that are inessential. Without a clear purpose, there are no criteria for deciding what is and is not essential.

So, if you wish to show someone what a tractor looks like, approximately, on the outside, you can show him a scale model. If you want to show him how it moves, you may push the model about. If you also wish to talk about it more generally, you can agree on the word 'tractor' to mean that sort of thing and you can vary the picture in your minds by exchanging instructions. If you wish to think about a red one, when your scale model was grey, you only have to say so. Thus a 'mental model' is much more flexible: it can, in fact, do things that no real tractor can do (such as travel at 100 mph). Mental models can be experimented on, therefore, but not systematically or in a disciplined fashion, because you cannot attend to all the detail, or the consequences (of your changed speed, for example), and you cannot remember all that was done.

To be really useful, therefore, except in the most imaginative of contexts, a model is better stated. A number of advantages immediately follow. Everything is explicitly stated, so there is never any doubt about the structure and content of the model. Once written (or drawn) it does not have to be remembered, because it can always be referred to, and it is clear to anyone who looks at it: it cannot be different things in the minds of different people. And so one could go on, but some of these advantages are not gained quite so easily. Suppose I describe a model by a diagram, as shown in Fig. 2.3. It may be obvious that it represents two circles (solid?) connected

FIG. 2.3.   A simple diagrammatic model.

to each other by a curved bar but, without further information, it could be two pipes viewed from one end, a weightlifter's bar or a pair of spectacles (see Fig. 2.4).

Of course, diagrams can and should be labelled and some indication given of what they are intended to represent. Nevertheless, it is extremely helpful if the viewer can immediately identify the meaning of any symbol used in a diagram. Lines to represent connections and arrows to represent directions are the most obvious and elementary conventions that we all accept and it is usual to extend these to include a whole range of components and processes. This is so, for instance, for representing electrical circuits, engineering diagrams and architects' plans, but it is also true in biology and agriculture. Flow diagrams, for example, are most simply drawn in the 'boxes and arrows' form shown in Fig. 2.5. These imply that material (or information) flows along the lines, in the direction of the arrows, from one box to another. Since the amount leaving one box and

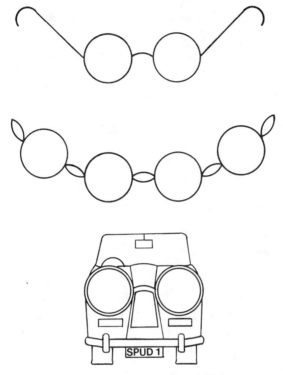

Fig. 2.4.   Possible derivations of Fig. 2.3.

arriving at the other depends upon the *rate* of the flow, this is commonly indicated as well. This has the advantage that the main factors affecting the rate can also be shown.

The drawing of such diagrams is therefore a matter of considerable importance and involves substantial skills. Final versions can be polished and made very attractive and this may best be done by someone good at such work, but the initial stages have to be done by the person whose thinking is being represented.

It is a salutary challenge to try and represent in a diagram one's knowledge about something, for a particular purpose. The last bit is important because it turns out to be virtually impossible to produce a picture of anything that includes all one's knowledge about it. If we know next to nothing about a gecko, for example, we cannot produce a picture at all. But if we take a familiar animal, such as a cow, superficially there

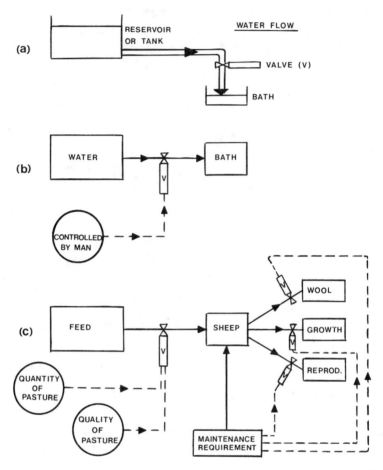

FIG. 2.5.   Flow diagrams. The flow could be of energy, for example, when all the boxes would represent quantities of energy in different forms. It is always worth considering whether such a diagram could be improved: this immediately demonstrates how diagrams help in thinking about how the system really works.

appears to be no problem in producing a picture. Figure 2.6 is an illustration of the most common kind of British cow and it is clearly recognisable because we already know about such cows and the picture contains enough information to identify the animal. A moment's thought, however, reveals that it is a picture of one side of a stationary cow. Yet we know much more than this. We know it has other sides and that

FIG. 2.6.

a great deal goes on inside; we know that it eats and breathes and moves about and we know that it reproduces and gives milk. None of these features appears in the picture and it is quite impossible to envisage a picture that did contain all that we know.

For any particular purpose, however, it is much easier to imagine a picture that contains all we need to know and it is better done as a diagram because it is clearer and less ambiguous. There are, nevertheless, many systems or organisms about which we know (and need to know) much more, even for a restricted purpose, than can be shown in a diagram. Furthermore, we may wish to represent processes quantitatively and dynamically over time.

It is this kind of complexity that is best dealt with by mathematical modelling in which the models are entirely abstract and expressed as equations of one sort or another. Amongst the advantages of this kind of modelling is that the calculating capacity and speed of computers can be harnessed to operate the model and to try out all kinds of changes on it. This experimentation on the model is very useful but its value depends, of course, on the validity of the model.

There are two ways of assessing the value and usefulness of a model but it is important to make this assessment in relation to the purpose for which it was constructed. Obviously, if the purpose of building a model was to determine gaps in the information available, then it is no use complaining about the model's predictive powers. The two kinds of assessment both relate to the purpose, therefore; first, related to whether or how well the model achieves its purpose and, secondly, to the 'correctness' of the way in which it does so. If a model is being used to simulate a real situation and it always behaves in the same way, in the sense that it gives the same answer or responds in the same way to changes in inputs or conditions, then the model is clearly useful for this purpose. However, it may do this for the wrong reasons, because things are correlated or linked in some way.

For example, a model could be constructed that associated crop growth with the application of fertiliser nitrogen when, in fact, the real mechanism might involve a deficiency in some minor element that was supplied as an impurity in the fertiliser. As long as the same conditions obtained, the same results would follow from applying fertiliser and the model would give correct predictions. So the model would be 'valid' for this purpose but it could not be 'verified' because the mechanisms assumed to operate would be found to be irrelevant and the assumptions false.

Models, of course, can relate to whole systems, or to parts of systems that are then regarded as systems in their own right, or to sub-systems.

SUB-SYSTEMS

It is highly desirable to use 'sub-systems' to mean something distinguishable (a) from systems and (b) from components of systems. As with the word 'systems' itself, we should be able to use the definition of 'sub-system' to distinguish the things that *are* sub-systems from the things that are not.

The main feature turns out to be the degree of independence. For example, if we look at a dairy farm as a system, we have no difficulty in recognising that each cow is a component and could be taken out and viewed as a separate system. Thus every system could, theoretically, be a component of another, larger system. It is also clear, however, that many studies could be made of one cow without necessarily learning anything about the dairy farm system. Studies could, for example, be concerned with the cow's reactions to climatic conditions or to feeds that it would never encounter in the dairy farm system considered. *Any* dynamic, practical view of the cow would need to take *a* food supply into account, however, and would have to recognise certain outputs (see Fig. 2.7). Why not *all* inputs and outputs? Because they are only required if our purpose demands them and the practical view of a cow inevitably involves feeding it. But if it is a

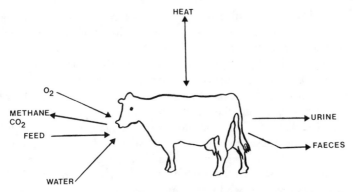

Fig. 2.7.   Minimum inputs and outputs in cow nutrition.

grazing cow, we cannot ignore the fact that its faeces may influence how much grass grows and how much of this grass the cow will eat. So we cannot relate our individual cow to the dairy farm system without enlarging our view at least to include these effects (see Fig. 2.8). The question is, how far does one have to go in this expansion and elaboration in order to obtain a part of the system that is nevertheless entirely relevant to it?

It is clearly extremely difficult to answer such a question without a picture of the whole system and it seems likely that sub-systems have to be extracted from systems; they cannot be built up independently. We then have two closely linked problems: how to describe systems and how to extract sub-systems from them.

We have talked about diagrams leading to mathematical models but this could apply to systems and/or sub-systems. Just as there are many

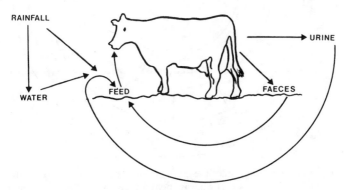

Fig. 2.8.   Minimum inputs and outputs for a grazing cow.

advantages in drawing a special diagram for a special purpose, so it may be better to draw one specifically designed to allow the identification and extraction of sub-systems.

One way of doing this is by circular diagrams. The idea is quite simple. It starts with putting in the centre of the diagram the output of central interest. The major factors that influence this output (which may be a ratio) are then grouped in a ring about this centre, with appropriate arrows pointing inwards. However, there may be effects of these factors on each other, on the same ring, and these can be indicated most neatly by circular arrows (see Fig. 2.9). Similarly, the items on this circle are influenced by other factors, which can be arranged to form the next circle, and so on (Fig. 2.10).

In this way, a picture can be built up systematically, always oriented to the central purpose and only expanded to include what appears to be

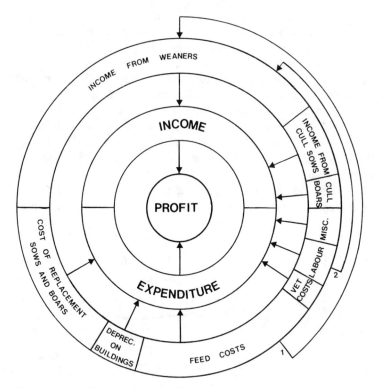

FIG. 2.9.   A circular diagram of the main factors determining profit in pig weaner production (Circular arrows: 1, what is spent on feed may affect performance and therefore profit; 2, the same may be said of labour costs).

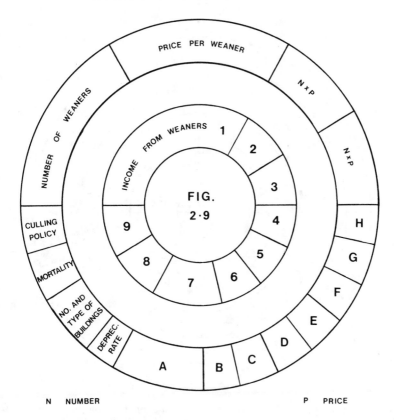

FIG. 2.10. An extension of Fig. 2.9 to include technical and biological factors.

A. Amount of feed for each class of pig (sow, boar, weaner).
B. Unit cost of each type of feed.
C. Number of visits × cost per visit.
D. Quantity of medicines × cost per unit.
E.
F. } Labour man/days × cost per day.

essential at each stage. This methodical elaboration of the pictorial statement applies to all the interactions between components, as well as to the components themselves. The process can continue until sufficient detail has been included, or until some kind of boundary is encountered. A simple example of the latter is shown in Fig. 2.11. In this case, the price of wool is assumed to be unaffected by the operation of the wool-producing process on the one farm considered and this component (W.P.) is therefore fixed as

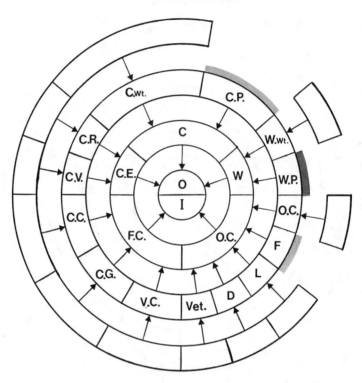

Fig. 2.11. A circular diagram of sheep production to show how firm (dark shading) or tentative (light shading) boundaries may be defined, in terms of the independence of the factor (e.g., W.P.) in relation to the effect of any part of the production system. (C.V. is affected, for example, by C.R. and C.P. may be influenced by C.wt.).

O	= Cash output.	C.P.	= Price per unit weight of lamb carcasses.
I	= Cash input.		
C.E.	= Value of cull ewes.	W.Wt.	= Weight of wool.
C	= Value of lamb carcasses.	W.P.	= Wool price.
W	= Value of wool output.	C.R.	= Culling rate.
F.C.	= Feed costs.	C.V.	= Value of a cull ewe.
O.C.	= Other costs.	Vet.	= Veterinary costs.
C.C.	= Cost of conserved forage.	D	= Flock depreciation.
C.G.	= Cost of grazed forage.	L	= Labour costs.
V.C.	= Value of concentrates fed.	F	= Fixed costs.
C.Wt.	= Weight of lamb carcasses.		

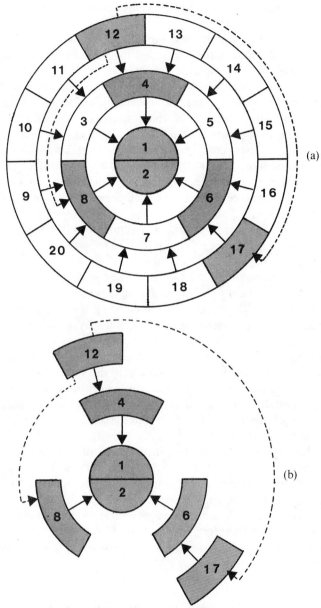

FIG. 2.12. The identification of a sub-system concerned with the effect of component number 12.

(a) The sub-system identified by shading.
(b) The sub-system extracted.

far as this system is concerned (and fixed by external influences). To simplify the examination of systems, it is always possible to 'fix' such values, or a range of them, for specific purposes.

Once the diagram is completed, it can be checked by others and any part of it amended as a result of suggestions and discussion that use the diagram as an unambiguous statement of what a system consists of, in relation to a specified purpose. Any amendment can then be equally clearly incorporated, so that a complex statement can be elaborated by people who only know about one part of it. This is quite different, incidentally, from the way in which mental pictures are changed in discussions from which each individual may emerge with a somewhat different view from that held by all the others.

The next step is to use such a diagram for the identification and extraction of sub-systems. For this purpose a sub-system can be specifically described as being concerned with the effect of changes in one component on the central output (Fig. 2.12(a)). The sub-system then consists of the centre, the component to be varied and all the components that are linked between these two by effects and interactions. If it is imagined that the components in a circular diagram are joined by wires to represent the arrows, then a sub-system can be 'lifted out' by taking hold of the centre and the selected component and extracting all that joins them (Fig. 2.12(b)).

The argument is that nothing else within the system affects the way in which the selected component influences the centre. A sub-system of this kind can therefore be studied on its own, with some assurance that the results of such studies will be relevant to the whole system from which the sub-system was derived.

In general, systems that justify substantial study have to be representative of a class of sufficiently similar systems, to which the results will also apply. This implies that agricultural systems have been classified, just as plants and animals have. Although this is not yet so, in fact, classification is just as essential for agricultural systems, and for exactly the same reasons (see Chapter 7).

Before leaving the subject of models and modelling it is necessary to deal with how they are tested.

THE TESTING OF MODELS

The usefulness of models depends upon their being relevant to the needs of the user and this can only be so if they are 'valid'.

Validation is simply concerned with establishing whether a model behaves sufficiently similarly to the real system being modelled.

The model represents a hypothesis that, for certain purposes, the real system can be represented in that way and validation is a test of that hypothesis. It is only necessary, therefore, to demonstrate that the model gives the same output (or whatever is being assessed) as the real system, over the same range of variables, controllable and uncontrollable.

However, two conditions need to be satisfied: one, that the model is tested against data not used in its construction; and, two, that what is meant by 'the same' output is specified in advance. In other words, the precision and accuracy required of the model must be specified in advance, recognising that the real system performance may vary considerably.

Until a model has been validated, it cannot be used with any great confidence, although it can be used to generate other hypotheses based on the assumption that it *is* valid.

If the validation procedure shows the model to be unsatisfactory, it is necessary to have measured many aspects of its performance in order to pinpoint where it is wrong. This means assessing separate processes within the model and is similar to verification.

Verification is the process of establishing the truth or accuracy of the processes and interactions on which the model is based and which are represented by equations in the model.

Such testing usually requires the same kind of experimental procedures as are used in science.

THE PRACTICAL VALUE OF A SYSTEMS APPROACH

So far, the value of a systems approach has been implied on largely theoretical grounds and the case may sound logically convincing (as I believe it is, but this should not be accepted uncritically, of course).

However, the approach has to be justified on grounds of practical relevance as well as on the basis that it helps us to think clearly about the subject.

Let us, then, consider how agriculture may be improved.

First of all, someone has to have an idea that he thinks or believes would result in practical improvement. If he has no evidence to support the idea, it is still possible to explore its likely effect, provided that we have a good enough picture of the agricultural system or component that we wish to change. Nevertheless, this is still a theoretical exercise.

Suppose, then, that this idea is backed up by observations from practice or experiment. The difference (see Chapter 6) is that experiments are carried out under more controlled conditions and their results are somewhat easier to interpret. In practice, it is difficult to connect an outcome with anything that was done because a lot of other things probably changed as well.

Thus the observation that the number of lambs born per ewe was higher after feeding extra minerals to the ram probably does not indicate a causal relationship. It might also have been a very wet summer, or less fertiliser might have been applied, or the stocking rate might have been less, or more rams might have been used: previous knowledge would suggest that the ewes were probably in better condition when they were mated and it is usually only in experiments that such a factor can be isolated.

However, since the experiment will look different from the farm situation, one is left wondering whether the results will apply to the latter. In other words, someone has to say what agricultural situations (or systems) it is thought that the results *will* apply to and to argue that the experimental situation represented a sufficiently similar system to make this likely.

The issue becomes even clearer if we consider how we would copy an improved system.

If we are shown a better (e.g. more profitable) system, whether in practice or at a research centre, and wish to adopt it, we have to know which bits to copy. We need to know whether we have to have the *same* cattle that were used (or just the same breed, age, weight and so on), whether the particular stocking rate, pasture species, fertiliser input, design of gates, method of ear-tagging, were all essential, whether the slope of the ground or the tree in the corner of the field were vital or trivial, whether we need to have the very man who operated the system successfully, and answers to a vast number of other questions.

Clearly we hope that the operator (or designer) will be able to tell us what the essential features are, so that we may copy only those. He may not be able to be sure about this, of course, but he may say that he has repeated his findings in several years and on several different fields, that he has used different cowmen and different cows. All this adds confidence, chiefly because he has developed a clear picture of the essentials (this is the same thing as the models mentioned earlier).

But can we be sure that he is right? One other feature that should add greatly to our confidence is if he can also say: I have put all these essentials together in a mathematical model, given them the relevant values (for stocking rate, fertiliser application, milk yield, etc.) and calculated the results for different weather conditions, and the answers come out near enough to those I actually get when I try it in practice. This tells us that the results of combining what he has identified as the essentials do appear to agree with results actually achieved in practice. Such calculations about complicated systems would not be possible without modelling techniques and the use of computers, which is one of the reasons why it is now possible to use a different approach. Two things of great importance to practical improvement flow from a systems approach, therefore. The first is the recognition that it is necessary to identify and describe the system studied and those to which the improvement will apply. The second is that it is possible to make calculations about the outcome as a basis for confidence in the result of the proposed improvement, and not only in one year but over a series of different years with varying weather and changing costs and prices.

Now, of course, this kind of thinking could be attempted by a specialist in any one discipline, but it would not relate to the whole system and it is this that characterises the systems approach. Consider the following example, taken from an attempt to improve livestock production—in this case, to produce a 'better' cow.

A 'BETTER' COW

An animal production specialist may look at a rather primitive milk production system in Africa and, viewing the single small Zebu cow producing a very low milk yield, may understandably consider that surely a 'better' cow could be provided.

This quite natural thought has often been the starting point for a selection programme, a cross-breeding programme or the importation of exotic breeds, usually resulting in a cow that was both genetically capable of a higher milk yield and also larger in size. This immediately suggests that any increase in milk output is likely to involve an increase in the feed supply—yet the latter is not necessarily considered in detail by the animal scientist, particularly if the feed comes from other parts of the farm or from by-products of crop production.

Thus an animal scientist may try to improve an agricultural system by 'improving' the animal component, without realising that the role of the animal may not just be, say, milk production. The animal may also carry out the cultivations for crop production and produce manure for fertiliser or fuel. In addition, the animal has to live and produce or perform in a

FIG. 2.13   An unimproved zebu cow producing a poor milk yield.

FIG. 2.14   A 'better' cow (Friesian), genetically capable of a much higher yield.

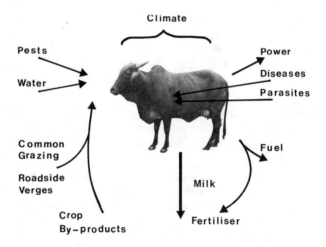

FIG. 2.15   An indication of the environment within which the zebu cow actually produced its yield.

climatic environment to which it is closely adapted—in ways that an 'improved' animal may not be. It may also be adapted to the local pests, parasites and water supply and, above all, be able to live on the available feed supply, much of which may be derived from crop by-products.

Similarly, a crop specialist might be tempted to 'improve' the crop

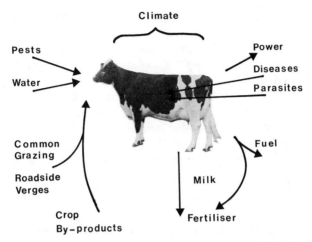

FIG. 2.16   The 'better' cow in the same environment as that of Fig. 2.15 with an indication of the milk yield it may be capable of if it survives.

component in directions that maximise the proportion of human food and reduce the proportion of by-products on which the animals are fed (or which are used for fuel, construction purposes etc.).

All this is illustrated in the sequence shown in Figs 2.13–2.16, for the substitution of a 'better' cow in a small-scale, milk production system. The 'better' cow may be genetically capable of greater milk production but, under the conditions of feed supply and environment in which it has to perform, it may actually produce *less* milk. This is quite possible if, as is likely, the animal is larger, with a greater maintenance requirement, and has to live on the same quantity and quality of feed. If the feed supply also has to be improved, it is clear that animal science is not the only discipline involved.

Equally, if the main effort is directed to improvement of the feed supply, it cannot be achieved solely by crop scientists, if only because the cultivations may depend upon available animal power, which is, in turn, influenced by the number, size and strength of the livestock.

It should also be clear that crop production has to achieve a balance between feeding livestock, feeding the farmer and his family and selling surpluses to generate income. With no income, there can be no inputs and this imposes severe limitations on the viability of a farm or a family. Yet it must be obvious that the balance to be achieved will depend upon the size and structure of the family, of the animals kept and on the markets for the sale of produce.

Such complex interactions are characteristic of agriculture and are most marked in the agricultural systems of developing countries. Indeed, the interactions frequently go well beyond the agricultural system and involve water and fuel collection, competing for time and labour. This serves to emphasise the need for a multidisciplinary approach and this can rarely be adequately embodied in any one individual.

So far we have been dealing with the systems approach in principle. In practice, it has been applied in several different forms by different groups of people. Since this has created some confusion, the essence of the main forms is given here.

FARMING SYSTEMS RESEARCH (FSR)

This has been variously considered as being the same as 'a systems approach' or in opposition to it. It has further been divided into at least three sub-categories.

The following descriptions are taken from a review by N. W. Simmonds (1985).

> *FSR sensu stricto*, is the study of farming systems *per se*, as they exist; typically, the analysis goes deep (technically and socio-economically) and the object is academic or scholarly rather than practical; the view taken is nominally 'holistic' and numerical system modelling is a fairly natural outcome if a holistic approach is claimed.
>
> *New farming systems development* (NFSD) takes as its starting point the view that many tropical farming systems are already so stressed that radical restructuring rather than step-wise change is necessary; the invention, testing and exploitation of new systems is therefore the object.
>
> *On-farm research with farming systems perspective* (OFR/FSP) is a practical adjunct to agricultural research which starts from the precept that only 'farmer experience' can reveal to the researcher what farmers really need; typically, the OFR/FSP process isolates a subsystem of the whole farm, studies it in just sufficient depth (no more) to gain the necessary FSP and proceeds as quickly as possible to experiments on-farm, with farmers' collaboration; there is an implicit assumption that step-wise change in an economically favourable direction is possible and worth seeking.

OTHER VIEWS OF A SYSTEMS APPROACH

There are, of course, many different views as to the way in which systems thinking should be applied.

One important difference, derived from the work of Peter Checkland, is that between 'Hard' and 'Soft' systems.

'*Hard*' *systems* are those involving industrial plants characterised by easy-to-define objectives, clearly defined decision-taking procedures and quantitative measures of performance.

Such systems tend to the mechanical, although biological systems are often of this kind (e.g. those represented by a single individual). Highly developed agricultural systems (e.g. battery hens) are also at this end of the range.

The more intimately people are involved as part of a system, however, the less appropriate this view becomes.

'*Soft*' *systems* are, by contrast, those in which objectives are hard to define, decision-taking is uncertain, measures of performance are at best qualitative and human behaviour is irrational.

In 'hard' systems, it is possible to focus on problems and endeavour to find solutions. In 'soft' systems, matters are rarely, if ever, quite so straightforward. In both kinds of system, however, it is still the case that 'improvement' is sought.

The distinction can readily be appreciated by focusing on extreme examples.

A bicycle is a mechanical system and, if it suffers from a puncture, it is quite possible to seek a 'solution' by repairing the puncture.

The problems of starvation in Ethiopia, on the other hand, have many dimensions (biological, social, military, climatic, etc.) and it is naive to seek a 'solution'. At the same time, there are obligations to help and this will turn out to mean trying to make changes that will move matters in a better direction, such that the result can be regarded as an improvement.

Of course, it is important not to over-simplify the question of improvement: what is an improvement for one person, may not be so for his neighbour. That is why two questions have to be posed at the beginning of any attempt to apply a systems approach. These are:

(a)   what is the system to be improved?

and

(b)   what constitutes an improvement?

Improvement cannot be sought by any method until these two questions have been answered and neither is simple. Indeed, both pose great difficulty and, in some circumstances, they cannot be answered to everybody's satisfaction.

The first requires a description of the system to be improved, in terms of its essential components, interactions and processes, boundary, inputs and outputs and, preferably, in such a way that possible improvements can be examined theoretically. In other words, a model is required, upon which experiments can be carried out. This model should be as simple as will serve the purpose and the latter is to help in determining potential improvements.

Until the second question has also been answered, therefore, it is very difficult to build the model. The only criteria available for deciding what must be included in a model and what should be left out, are the objectives and purpose of the model and of the activity or system being modelled.

If, for example, increased profit is the objective of improvement, then the model must be built in economic terms and contain all the factors relevant to the formation of profit.

It is often assumed that profit *is* the most likely objective and, indeed, that the aim is usually to maximise it. But in many agricultural systems, especially in developing countries, profit is not the most important objective. Stability of output or profit over a long period may be more important; or the objective may not even be expressed in monetary terms at all.

Very rarely, in fact, does anyone wish to *maximise* profit, except within a great many constraints, even where an increase in profit is sought. Peace of mind, security, reduction in drudgery and the satisfaction of current need may all count for a great deal. In general, an improvement in efficiency is sought, but this can be expressed in many different ways.

In many circumstances, as with most other human aims, the objectives may be neither simple nor single and they may not remain constant over time. Difficult as it may be to agree on objectives (which may differ markedly between producer, employee, consumer and the nation), it is essential to examine the issue carefully before embarking on an improvement programme.

In general, improvement is achieved by one of three main methods:

(a) Advice on component changes;

(b) Adoption of innovation

and

(c) Copying.

(a) Advice

Many producers rely upon advice as a basis for action designed to improve their animal production systems. The sources of advice are numerous, some associated with government services, some with commercial concerns and some paid for from independent consultants.

It must be clear from what was said earlier that successful advisers must be able to relate their advice, which, of necessity, will usually concern only part of a system, to the functioning of the whole system.

Otherwise, the producer has to depend upon his own judgement and knowledge of the system he is operating, in order to decide whether the advice is sound for his system or which parts of it are relevant.

In either case, a picture of the whole system and the way it operates has to exist, if only in the mind of the farmer or of the adviser.

Advice need not be concerned with new knowledge or practices and, in most circumstances, is based on ideas that have already worked in practice, in other systems. Part of the judgement then required is whether the system now considered is sufficiently similar to those in which the idea has previously been applied successfully.

(b) Innovation

Improvement can also come from innovation, in the sense that new ideas are involved that have not been tested in practice before, although they may have emerged from research and development studies.

The pioneer farmers usually engaged in this kind of experiment tend to be interested in trying out new ideas—sometimes in originating them—and can often afford to take the risk of an innovation not working. Indeed, such farmers may be able to benefit by being in the van of progress when innovation is successful and thus able to survive the occasional setback.

In extreme cases, such trial and error can be carried out without any kind of systems approach, although it is hard to imagine successful farmers who do not have *some* picture of the system they are operating.

It is certainly important to realise that the picture or model required may be quite different (in content, detail, level of sophistication) according to whether the purpose is to operate, repair or invent new systems.

(c) Copying

Very often, practical improvement is achieved by copying what is done by others, whether they are other farmers or research workers.

The main problems of copying are of three kinds. First, there is a matter of confidence that the system to be copied is better than that already being operated. This has to be judged by the results that matter to the prospective copier and by his confidence that he has available all the information he needs (and that it is reliable). Judgement may also require evidence over a period of years, involving a range of weather conditions (as discussed earlier).

Second, there is a matter of relevance: there has to be good reason to suppose that the system to be copied is relevant to the climate, aspect, topography, soil type and so on, of the copier's farm.

Third, there is the question of exactly what to copy. Part of the difficulty is that some things *cannot* be copied. It is not possible to have the identical animals, the identical pasture or the identical staff; and all of these may be important and any one of them vital to the success of the system to be copied.

Another part of the difficulty is to determine which of the things that could be copied are important to the success of the system and which are not.

Related to this is the question of *how similar* the things that have to be copied have to be in order to ensure success. The similarity may relate to breed or strain of animal, level of feeding or fertiliser use, type of fencing used, availability of water, timing of mating, health care and veterinary treatment. Any one of these may matter greatly and success may depend upon getting them all right.

In general, one has to rely on the operator (of the system to be copied) to distinguish between what matters and what does not and to answer all the detailed queries. But, unfortunately, the operator of a successful system may not actually know these answers, even though he may think he does. Nor are the answers easily determined, even at a research institute.

One way of gaining confidence that the vital elements have been identified is to construct a model, as mentioned earlier in this chapter.

3

Ways of Looking at Agricultural Systems: The Problem of Description

In the last chapter, it was argued that it is essential to be able to describe a system in terms which make it possible for it to be copied, for example. The trouble is that one picture is never enough.

A photograph of grazing cattle and a set of financial accounts are both perfectly legitimate views of an agricultural activity. Neither tells the whole story but each illustrates one aspect much better than, let us say, a photograph of an accountant at work on the books or a verbal description of cattle at pasture. In order to get a picture of a whole agricultural system, it is necessary to have a large number of such pictures, simply because there are a great many aspects to be considered. Although this raises problems and appears complicated, this multi-facetted feature is one of the attractions of agriculture. At the same time, it has to be recognised that most people would say exactly the same about their own (different) subjects, which emphasises again that complexity is really in the mind of the observer who is free to adopt as many viewpoints as he wishes.

One general proposition can usefully be restated at this point: there is no possibility of one picture (or model) representing all these possible viewpoints. Attempts to produce such pictures are doomed from the start. It may be possible to construct a multi-purpose model but it is not possible to construct one for *all* purposes. Furthermore, it is usually much better to build a model or make a picture for each specific purpose or to illustrate one particular aspect.

That this is so in ordinary common sense terms can be seen by contemplating Fig. 3.1 and comparing it with Fig. 2.6.

We have to consider, therefore, what are the most useful ways of looking at agricultural systems and, of course, these will have to be related to specific purposes. If our purpose is to improve the profitability of a farm, for example, at least two views will be needed. The first will be concerned with money. This could be a budget (Table 3.1) or a cash flow diagram (Fig. 3.2) or it could take several other forms. The second would be a picture that included the biological processes involved, so that we could see what might be changed in order to increase profit (the circular diagrams used in Figs. 2.9, 2.10 and 2.11 are of this kind). It is vital in this to realise the interactions between components. To take a rather extreme example, it is no use economising by halving the feed fed to a pig, just because feed costs have

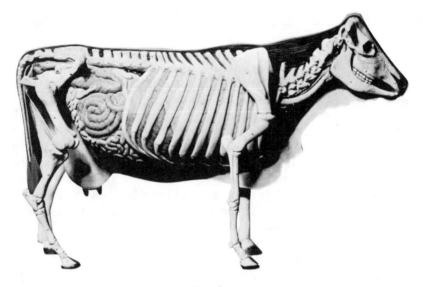

Fig. 3.1.

been identified as representing 70 % of the total costs. Very often, feed or fertiliser inputs have to be *increased* in order to improve profitability. A reduction in the mortality of piglets born seems a much more obvious improvement, but even here there may be additional costs that have to be taken into account.

However, if we are taking a regional or world view of the problem of feeding people, a picture that represents the production of dietary energy and protein may be more relevant. Such pictures can also take many forms (see Tables 3.2 and 3.3), including that of a flow diagram (Fig. 3.3). It would be a help in producing these views of agricultural systems if they could be sensibly related to a classification scheme so that we would be dealing with

TABLE 3.1

EXAMPLE OF CASH FLOW ON A 100 ha MIXED FARM (SEE FIG. 3.2)

(A complete budget would also include opening and closing valuations for livestock, buildings, materials in store, etc.)

	Jan.	Feb.	March	April	May	June	July	Aug.	Sept.	Oct.	Nov.	Dec.
Receipts												
Milk (60 cows)	4000	3500	2900	2000	1500	776				4000	4200	4400
Cattle (calves and culls)						660	660			825	825	825
Sheep and wool (80 ewes)				500	1000	800	1400					
Barley (50 ha)		8250		5250								
Pigs (20 sows)	1515			1515			1515			1515		
Contract		376										
Sub-total	5515	12126	2900	9265	2500	2236	3575			6340	5025	5225
Expenses												
Cattle ⎫ replacements									3500			
Sheep ⎬ replacements										700		
Pigs ⎭ replacements	200		60			200					2000	
Bought food: Cows	2000	1740										
Bought food: Sheep												
Bought food: Pigs	800	1880		800			800		2000	855	140	
Seed (corn and grass)	3375											
Fertiliser												
Wages	580	580	580	580	580	580	580	580	580	580	580	580
Rent	1250			1250			1250			1250		
Machinery	2500			2500			2500			2500		
Sundries	375	375	375	375	375	375	375	375	375	375	375	375
Sub-total	11080	4575	1015	5505	955	1155	5505	955	6455	6260	3095	955
Monthly balance	−5565	7551	1885	3760	1545	1081	−1930	−955	−6455	80	1930	4270
Cumulative balance	−5565	1986	3871	7631	9176	10257	8327	7372	917	997	2927	7197

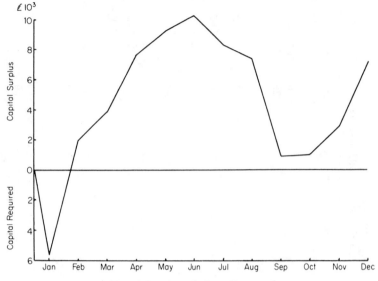

FIG. 3.2.   A cash flow diagram.

the main kinds of agricultural system. The idea has not been fully worked out yet by anyone, but is further discussed in Chapter 6.

However, it is possible to consider what the most important views are likely to be by considering what are the most important purposes for which we would require them.

TABLE 3.2

WORLD CONSUMPTION OF DIETARY ENERGY
(% distribution by food categories and regions)
(after Brown and Finsterbusch, 1972)

Regions	Main crop products (Cereals, roots, tubers, fruits, nuts and vegetables)	Sugar	Fats and oils	Animal products
North America	33·5	15·8	19·9	30·8
Western Europe	50·3	11·2	16·8	21·7
South America	63·0	14·0	8·0	15·0
Africa	81·6	4·1	7·5	6·8
Asia	85·9	4·1	5·3	4·7

TABLE 3.3
WORLD CONSUMPTION OF DIETARY PROTEIN
(After Duckham *et al.*, 1976)

| | Protein consumption | |
	Total ($10^6 t/y$)	Per head (g/day)
World	87·4	66
Developed countries	23·9	89
Developing countries	36·7	55
Centrally planned economies	29·2	67

It usually turns out that the main purposes are in one of the following categories:

(1)   Management of an agricultural system.
(2)   Repair—when something goes wrong.
(3)   Improvement.
(4)   Innovation.

In practical terms, farms are operated by managers, who may or may not be owners, and these managers may call in advisers or consultants to help them with overall management or with particular aspects, down to quite fine detail (e.g. how best to sow a particular crop or adjust a milking machine). If things go wrong, other specialists are called in, such as plant pathologists or veterinarians, in order to repair some part of the system and restore normal functioning.

Improvement and innovation really only differ in the extent of the change that is contemplated.

Pictures therefore tend to be concerned with either (a) monitoring or (b) contemplating the consequences of change.

(a)   MONITORING

Monitoring is rather like taking someone's temperature: it is assumed that all is well as long as the value monitored is within some satisfactory, expected range. If it is not, it indicates that something is wrong but it may be a long and difficult process to diagnose exactly what the problem is.

Monitoring of agricultural systems may take many forms but the most

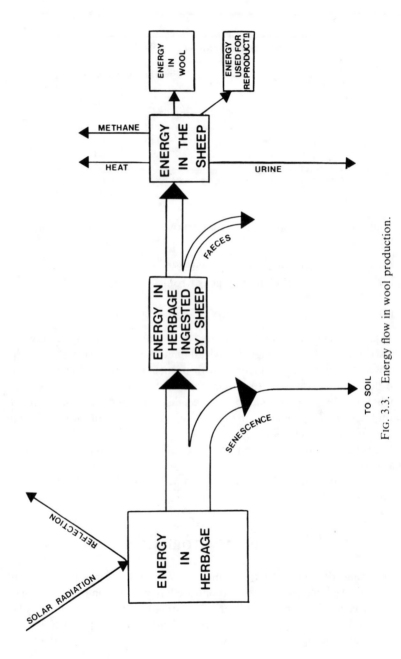

FIG. 3.3. Energy flow in wool production.

usual are concerned with financial monitoring, checking actual against projected budgets and the monitoring of physical performance, such as crop or cow yields, animal growth rate or seedling emergence, to see whether it matches up to targets set. In many cases, none of this involves a picture of how the system works although a great deal of the information collected by monitoring is used as a basis for corrective action, within the general task of management. This is an activity rather like driving a car, in which monitored information about speed, fuel supply, engine temperature and oil pressure may be used as a basis for further acceleration, braking or refuelling. In fact, a picture of the system exists, relating these actions with their probable consequences, but it is a relatively simple one: nevertheless, this may be all that is required.

If the system goes wrong, however, a much more detailed picture is required in order to know what to do about it.

(b) THE CONSEQUENCES OF CHANGE

Even more detailed pictures are needed if we wish to change part of the system or the level of an input, and the bigger the change contemplated, the greater the requirement for detailed knowledge of internal workings and relationships.

The sort of questions that one would wish to pose of an agricultural system include: What governs the output? What is the effect of altering the nature or quantity of an input? And, most important: What is the efficiency of the system and how is this influenced by change?

The last kind of question is the most important because agriculture is always about efficiency. This idea often raises objections because it is immediately (but erroneously) supposed to imply that particular (e.g. financial) efficiencies should be maximised, whatever the consequences to pollution, amenity, way of life and so on. This is hardly ever considered seriously by anyone, however, as a major objective. In fact, maximising one efficiency usually reduces the efficiency with which other resources are used. Economic efficiency tries to cover all the main inputs and outputs but there are always some that cannot be expressed in simple financial terms and constraints are therefore accepted. A common example of this would relate to how many hours the farmer himself works and whether he and his family have a holiday.

In agricultural operations, therefore, there is characteristically a considerable balancing exercise between different aspects of efficiency and it would normally be expected that the final outcome will be a compromise.

This still means that efficiency has to be sought wherever it is possible but that the effects on other aspects of systems operation have to be considered. This means that, although it will be necessary to consider separately the efficiency with which each resource is used, it is also necessary to consider the effect of any change in terms of overall efficiency.

If we mean by efficiency simply the ratio of output/input, it is clear that we can include whatever we wish in each half of the ratio and we can express each in several different ways (e.g. money, energy, nitrogen) according to our interest. Nor do we have to express each half in the same terms, since we are often interested in such ratios as milk output per cow, growth rate per day, grain yield per unit of irrigation water applied or cash output per man.

Having chosen an appropriate ratio, it can be calculated for actual or postulated circumstances and the outcome judged as satisfactory or not. If the answer is *not* satisfactory, however, it is important to know what can be done about it.

There are, of course, two ways of improving a ratio: one is to increase the numerator at the top and the other is to decrease the denominator. But this does not tell one how to do either or what the consequences might be. A simple expansion of such ratios (see Fig. 3.4) indicates the main factors contributing to the ratio and the fact that the top and bottom parts of the ratio are not independent. Thus, although the efficiency of feed conversion by animals is extremely important (where the ratio is output per unit of feed consumed), it cannot necessarily be improved by reducing the amount fed. A superficial examination of the equivalent monetary ratio (cash output of product/cost of feed input) suggests that improvement could be achieved by (a) an increase in output or the price received per unit of it, (b) a reduction of cost per unit of feed or (c) a reduction in the amount fed. The fact is, of course, that more output may require more feed and less feed may lead to lower output.

These sorts of interaction are of exactly the same kind as were represented in the circular diagrams in the previous chapter (e.g. Fig. 2.9) and it is in this way that the expansion of ratios leads to diagrams of the kind described.

An approach that starts with some measure of efficiency thus also arrives at a need to describe the way the system works in order to find out in what ways efficiency could be improved or how it would be changed by proposed modifications to the system and its inputs.

This means that whatever the value, or necessity, of looking at

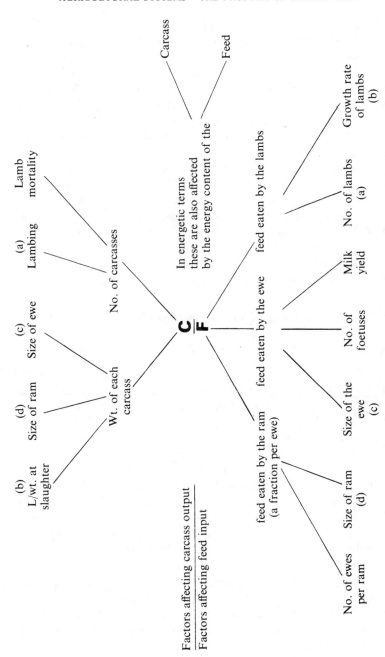

FIG. 3.4. The expansion of an efficiency ratio. In this example, energetic efficiency of feed use by sheep (over 1 year) is calculated as Carcass energy output (C)/Feed energy input (F) per ewe (letters in brackets indicate the more obvious connections).

agricultural systems from an economic point of view, it is frequently necessary to look also at the underlying biological mechanisms if there is any question of changing parts of the systems.

Thus, although biological and economic efficiencies are different, and will be considered separately in Chapters 4 and 5, it is important to be prepared to mix the two wherever this is appropriate. Indeed, the criterion for including something in one's picture of a system should be whether or not it influences the output rather than whether it falls within one discipline (subject) or another.

Increasingly, in developed countries, it is becoming necessary to take a wider view of agriculture, not only in terms of its effects on the animals used, the people involved and the health of the consumer of its products, but also on the environment as a whole, its fauna and flora, its appearance and the people who visit it. These issues are further considered in Chapter 15. What needs to be recognised at this point is that future agricultural systems may need to take a different view of efficiency and agricultural use of the land may need to be integrated with other (non-agricultural) forms of land use.

THE MAIN SYSTEMS CONCEPTS

It may be helpful to summarise at this point the most important systems concepts, as follows:

(1) Systems are identifiable entities with important properties and attributes that are quite distinct from those of non-systems.

(2) Systems can be of any size or complexity, from a molecule to an elephant or a universe, and any system can be either a component or a sub-system of another system.

(3) When a system is a component of another system it has specific inputs and outputs resulting from its interactions with the rest of the system: an independent system is not bound by such interactions.

(4) In practice, no system is completely closed or completely independent but the differences between an object as a component and as a system are usually very large.

(5) All systems can be modelled but it is not always practicable to construct a model at any given level of detail.

(6) The level of detail and the structure of the model should be related to the purpose of construction and should be as simple as will serve the purpose.

(7) Models should always be capable of validation and, where possible, should be validated.

(8) The required precision and accuracy of a model should be specified in advance: otherwise validation is not possible.

(9) Where systems are too large or too complex to be studied in their entirety, sub-systems may be identified that can usefully be studied separately.

(10) Sub-systems have a degree of integrity and independence of the whole system, such that they can be studied separately and the results of such studies incorporated into a model of the whole.

(11) The independence of sub-systems depends on the existence of only a few main interactions with other sub-systems or components of the whole system.

(12) Sub-systems will usually have the same output as the main system but relate to only some of the components and thus to only some of the inputs.

(13) Improvement (or indeed any response) of a system, due to changes in one or more components, cannot be predicted without the use of a model, of some kind, representing that system.

4

Biological Efficiency in Agriculture

Although agriculture involves a great deal more than biology, it is nevertheless based on biological processes. It is impossible to conceive of an agriculture or an agricultural system that is not based on one or more biological processes. Indeed, a part of the definition proposed in Chapter 1 was the controlled use of animals and plants.

When Man was a food gatherer and a hunter, the biology of the plants and animals that he harvested remained largely unaffected by his activities. Man was just one more consumer or predator and, unless his population or his activities got out of hand (as with the slaughter of the North American bison), the main biological processes continued normally.

Meat would have been obtained but not milk, presumably, since this demands the kind of control that is characteristic of agriculture. In the case of crops, control takes the form of soil cultivation, sowing, fertilising and protection from pests, diseases and animals and from competition from weeds. With animals, in addition to protection, there is also the provision of feed and water and some degree of restraint. Amongst the methods used to make animals less dangerous and more tractable, castration of the male has been used from quite early times. Some degree of control of reproduction is also a feature of animal production.

Many of these control methods involve manipulation of the original biological processes but others have the effect of eliminating some of them. The aim, of course, was to remove processes that were inimical to the one directly involved in producing whatever was required. However, some of the changes that occurred not only eliminated undesirable processes but removed or restricted parts of the main production process, in order to substitute something else, generally better or more easily controlled. Artificial insemination is a clear, relatively recent example. Semen can be

52

collected from a selected few males (of above average quality), stored until required (at least for some species, such as cattle), diluted to standard strengths and inserted into known cows at chosen times. This is all safer and more efficient than the natural process, which is thereby eliminated. Incidentally, this has allowed the development of breeds of turkey that are unable to mate normally and *have* to be inseminated artificially.

I am not arguing here about whether this is right or wrong, desirable or not, but simply noting the developments that have occurred. They have, in fact, been quite widespread and varied and many of them have had the effect of *reducing* the extent to which agriculture has a biological content.

Some of these changes are listed in Table 4.1. The battery hen is a good example, where the point is reached when anything that can be done for the bird *is* done for it and the bird itself does nothing that could be done for it. Thus feed and water are brought to it and faeces and eggs are removed as they are produced. Temperature, light and humidity are controlled, flies are killed, eggs are hatched and chicks reared under separate, controlled conditions. The biological system is reduced virtually to the bird itself and

TABLE 4.1

EXAMPLES OF THE DISPLACEMENT OF BIOLOGICAL PROCESSES IN AGRICULTURE

Biological process displaced	Non-biological process substituted
Incubation of eggs by hen bird	Electrically heated incubator
Natural service by male animal	Artificial insemination
Collection of feed by animal	Automated provision of processed feed
Grazing	'Zero-grazing' (cutting and carting of herbage)
Deposition of excreta on the land	Collection of excreta from housed animals → disposal, treatment, spreading on land
Natural control of pests and weeds	Use of pesticides and herbicides
Natural immunity to disease in animals	Use of vaccines
Natural hormonal processes	Control of light, daylength and temperature; use of synthetic hormones
Fixation of atmospheric N by bacteria	Manufacture and application of artificial nitrogenous fertilisers
Recycling of phosphorus and potash; extraction from soil by deep rooting plants	Application of artificial fertilisers
Water uptake by deep roots	Irrigation
Natural suckling (of lambs and calves)	Artificial rearing on milk substitutes

even within this there may be supplemental hormones introduced. Similar developments have occurred with other poultry, with pigs and, to a much lesser extent, with cattle, and even less, with sheep.

Indeed, the giant feedlots of North America for beef production on a grand scale take all the feed to the cattle and remove all the dung, involving considerable transportation. Accumulations of dung from large animal production enterprises have come to represent a major disposal problem and a potential pollution hazard.

The advantages of these developments in animal production have centred on the greater performance (milk yield, egg output, growth rate) achieved per unit of time, per man and per unit of feed (and thus per unit of land employed), as a result of better control of such things as diet, feed intake, parasites and disease.

The major additional costs have been machinery and fuel in the case of animal production, and, in addition, fertilisers, pesticides and herbicides in crop production. The advantages in the latter have also been better performance, in terms of harvestable yields per man and per unit of time and land.

Essentially, where control has been achieved by the elimination or reduction of biological processes it has involved the substitution of support energy. Of course, in many instances, such as weed control, it would be quite possible to use large amounts of labour rather than energy in the form of machinery, fuel or herbicides. It has been estimated that about one-third of the total support energy inputs goes to produce more food and about two-thirds to save labour. So it would be possible to reverse some part of this trend, without lowering food output, provided that labour was sufficiently cheap.

However, there are sometimes other possibilities. For example, biological control of pests and weeds is successfully practised in some cases. This involves the use of other organisms, mostly insects, to control those regarded as undesirable. Table 4.2 lists some of the successes in this field.

Why should alternative be sought? What is wrong with the present situation? The answers to these questions depend upon who is answering and from what point of view.

There are public concerns in many countries about the ways in which modern agriculture has developed. The main concerns are given in Table 4.3. The precise concern and the reasons for it vary with the location. Concern about animal welfare, fertilisers and agrochemicals tends to be much greater in countries that produce more than they need. That is not to say that developing countries should not be concerned with pollution and

TABLE 4.2
BIOLOGICAL CONTROL, EXAMPLES OF SUCCESS
(Sources: van den Bosch (1971); Edwards and Heath (1964); van Emden (1974);
DeBach (1964, 1974))

Organism controlled	Controlling organism	Location
Winter moth	Insects (a tachinid and ichneumonid)	Canada
Southern green stink bug	An egg parasite	Hawaii, USA
Olive scale	A chalcid ectoparasite	California, USA
Rhodesgrass scale	A wingless parasite	Texas, USA
Walnut aphid	A parasitic wasp	California, USA
Alfalfa weevil	An ichneumonid	Eastern USA
San Jose scale	A parasitic chalcid	Switzerland and California, USA
Cottony-cushion scale	A predatory beetle	California, USA
May beetle larvae	Giant Surinam toads	Puerto Rico
Apple sucker	A fungus	Canada
Diamond-back moth	Bacteria	USA
European spruce sawfly	A virus	Canada and Sweden
Coconut moth	A tachinid fly	Fiji
Woolly aphid	A parasitic wasp	UK
Glasshouse whitefly	A parasitic wasp	UK

TABLE 4.3
PUBLIC CONCERNS

(1) *Farming Methods*
    (a) Animal welfare
    (b) High inputs of fertiliser
    (c) Use of agro-chemicals
    (d) Destruction of hedges
    (e) Felling of trees
    (f) Drainage of wetlands
    (g) Monoculture
    (h) Straw burning
    (i) Offensive smells and sounds

(2) *Agricultural Products*
    (a) Nature of the produce re health, diet, etc.
    (b) Residues of chemicals
    (c) Additives

(3) *Economic*
    (a) Costs of agricultural support
    (b) Prices of food

similar matters but it is understandable that they may not receive a high priority. Drainage of wetlands looks quite different to a country short of food and to one that suffers from overproduction. Monoculture may worry one country because of the resulting appearance and another because of implications for soil erosion or pest control.

Concerns about diet vary in two main ways. One is simply related to how hungry you are and the other flows from whether the food is processed by the farmer or consumer or by a food industry. In countries such as the UK, some 70–80% of the food consumed is processed in one way or another, most of it passing through a sophisticated food industry. It is here that additives may be used, as preservatives or to add flavour or colour, for example. Economic concerns are usually present, in one form or another.

In most developed countries, agricultural support—in the form of subsidies or artificially maintained prices—is at a high and often intolerable level.

In many developing countries, in an endeavour to keep food prices down, the prices paid to farmers are too low to provide adequate returns, much less to provide incentives to produce more.

For all these reasons, alternative methods may be needed and, in Europe, for example, there has been a steady increase of interest in 'organic' farming. Of course, it is easy to ask what other kind of farming there is and to argue that ecological systems are designed to absorb inputs and adapt accordingly.

There is no doubt, however, that many people prefer to consume food produced with few (if any) chemicals and that organic methods do provide lessons that can be applied more widely.

Organic farming is defined in various ways.

DEFINITION OF ORGANIC FARMING

Although a rather confusing, and sometimes misleading, term, organic farming has come to mean both an attitude of mind and a set of farming practices.

Both are characterised by encouragement of favourable biological cycles and avoidance of 'chemical' fertilisers, pesticides, herbicides and animal feed additives.

The definition used by a USDA Report on Organic Farming (1980) was as follows:

'Organic farming is a production system which avoids or largely excludes the use of synthetically-compounded fertilisers, pesticides, growth regulators, and livestock feed additives. To the maximum extent feasible, organic farming systems rely upon crop rotations, crop residues, animal manures, legumes, green manures, off-farm organic wastes, mechanical cultivation, mineral-bearing rocks, and aspects of biological pest control to maintain soil productivity and tilth, to supply plant nutrients, and to control insects, weeds and other pests.'

One characteristic of organic farming is that it uses less 'support' energy—energy other than that derived from current solar radiation. Coal, gas, petrol and oil are regarded as non-renewable resources because, although derived from past vegetation and thus from solar radiation, the processes involved take a vast period of time. It is this time-scale which makes the difference. After all, solar energy fixed in mature trees takes quite a time to accumulate and a fifty-year-old man burning oak logs is consuming a resource which, in his lifetime, is non-renewable. Even so, the time involved is nothing like that required to make oil or coal. Not that there can be anything wrong with using up such stores: they benefit no-one if not used at all. But the rate at which they are used and the precise purposes for which they are employed need to be based on clear priorities if the supply is in danger of running out in the foreseeable future. Of course, other sources of energy (wind- or wave-power, nuclear fusion) may transform the situation, but until this can be guaranteed it will be necessary to consider the role of support energy in agriculture. This will be discussed further later in this chapter (and in Chapters 11 and 12); here I simply wish to establish the connection between biological processes in agriculture and the use of support energy.

Agriculture began as an attempt to use solar energy by exploiting the plant's ability to fix it during photosynthesis and by exploiting the animal's capacity for digesting fibrous plant material that is of little use to Man directly. The great virtue of agriculture, therefore, is that it can harness current solar radiation in the production of food and fibre. No other industry can do this to any worthwhile extent. If the present reserves of fossil fuel run out and no new methods of energy release have been found to replace them, we will all depend upon current solar radiation to an even greater extent and agriculture, as the main method of using solar radiation, will become even more important than it is.

Now the ability of agriculture to harness solar radiation directly depends upon photosynthesis and, not quite so directly, upon other biological processes including animal growth and reproduction.

The substitution of support energy for solar energy is thus somewhat antagonistic to the central function of agriculture and it seems worth

TABLE 4.4

SOLAR RADIATION RECEIVED ON EARTH

(after Holliday (1976))

Maximum daily value	$J\,ha^{-1}\,day^{-1}$
Total radiant energy available to crop	$1\,674 \times 10^8$
Total of photosynthetically active radiation (50%)	837×10^8
Total used by a crop canopy	652×10^8

The total radiation energy reaching the outer edge of the earth's atmosphere (the solar constant) $= c.\ 8.36$ $J\,cm^{-2}\,min^{-1}$.

considering whether we could *increase* the number of biological processes that could help us to use more of the enormous quantities of solar radiation that reach the earth each day (see Table 4.4).

BIOLOGICAL SYSTEMS IN AGRICULTURE

There are many biological systems *within* agriculture. Every animal and every crop can be regarded as a separate system and so can parts of animals (such as the rumen or fourth stomach of the cow), the pests of crop plants or the parasites of animals.

However, the importance of these components to agriculture rests on their role within agricultural systems and, without their connections to the rest of these systems, these components are unhelpfully isolated. Thus, as was suggested in Chapter 2, studying isolated components, without regard to the systems of which they form a part, will not necessarily lead to agricultural improvement. Biological systems, then, if they are to be relevant to agriculture, must represent sub-systems of whole agricultural systems or biological views of agriculture.

Let us consider first what a biological view of agriculture would look like. A generalised picture would have to contain growing plants and animals. Both would have to produce products (and by-products) and unusable portions (some of which might be recycled) and both would have to devote some of their output to reproducing the next generation. Water and nutrients would have to be shown as major inputs, as would the energy source for plants: some plants would be devoted entirely to feeding animals. These features seem to represent the minimum that a general picture should contain (see Fig. 4.1).

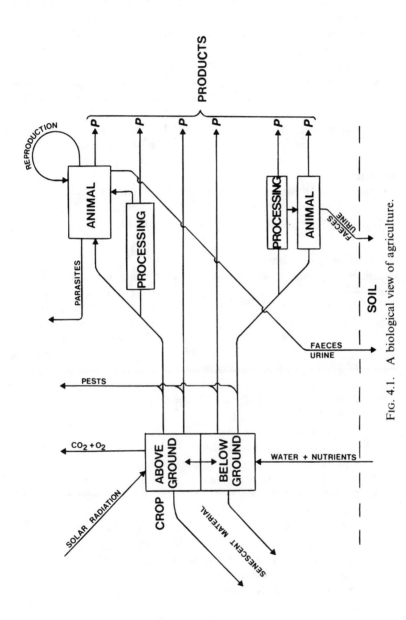

FIG. 4.1. A biological view of agriculture.

Partly because it is a general picture, it does not indicate particular plants and animals, or the rates of processes, or the environment within which the system operates. Thus, although soil is present and has a water content, it is neither waterlogged nor drought stricken: similarly, it is neither flat nor on a steep slope, neither smooth nor strewn with boulders. So, although herbage can be cut or grazed in a generalised system, some of the important factors that would determine the method of harvesting are not indicated at all.

As usual with models, they are more useful if they are quite specific and related to particular systems and processes. The latter are rather numerous, however, and it is difficult to select a few that may be regarded as characteristic.

Perhaps the most useful approach is to consider what are the characteristic *roles* of plants and animals in agriculture and use these as a basis for thinking about the essential biology of agriculture.

The Roles of Crop Plants

The main purposes for which *plants* are grown may be listed as follows:

(1) Amenity (a) to clothe land, as with lawns, to improve appearance,
         (b) to provide attractive views—distant, as with woods, or close, as with garden flowers—and odours,
         (c) to provide 'furnishings', such as cut flowers and other decorative material
and recreation (e.g. playing fields, nature trails).

(2) Production of food for direct consumption
    (a) from seeds, e.g. cereals
    (b) from roots, e.g. sugar beet and carrots
    (c) from tubers, e.g. potato
    (d) from fruits, e.g. apple, date, grape
    (e) from leaves, e.g. cabbage, lettuce

(3) Production of beverages that are not strictly foods, e.g. alcohol, tea, coffee, wine, maté.

(4) Production of oils (e.g. cottonseed, palm, olive), waxes, pectins, gums, resins, tannins, dyes and perfumes.

(5) Production of drugs and medicines (e.g. opium, quinine) and insecticides (e.g. pyrethrum).

(6) Production of fibre (a) soft fibres (e.g. flax)
         (b) hard fibres (e.g. sisal)
         (c) surface fibres (e.g. cotton)

(7) Production of timber (a) hardwoods (e.g. oak)
                       (b) softwoods (e.g. spruce)
(8) Production of feed for animals (a) herbage, such as grass
                                  (b) roots and forage crops
                                  (c) cereals and pulses
(9) Physical function *in situ*, such as provision of shelter belts or for erosion control.
(10) Increase in soil fertility, as with green manures that are ploughed in again and legumes that fix atmospheric nitrogen and contribute this to the soil via dying roots.

Some of these purposes are not really agricultural and the most important are probably (2), (7) and (8). The importance of particular crop plants varies from one place to another, of course, and to the individual farmer any one of them (e.g. opium or tea) might easily be the most important.

Two further roles of crop plants are becoming recognised as of such importance that they should now be separately identified.

(11) Production of raw materials for industry.

There is some overlap with (2) and (3) (because there is a food industry) and with (4), (5), (6), (7) and (8), because industries have developed to process these commodities.

However, a great deal of thought is now being given to the use of, for example, food products, such as cereals, sugar beet and potatoes, as sources of non-food products, such as starch and ethanol. This is not merely as a means of disposing of surpluses, in the EEC for example, but because the technology now exists for producing such crops very efficiently but in excess of the demand for them as food.

The potential for industrial use depends greatly on the cost of production and is thus influenced by the price of land and other aspects of the high cost structure currently found in the UK, for example.

(12) Production of fuel.

Here the situation varies greatly from one country to another. The majority of food is cooked and, in many countries, the fuel has to be produced locally, either on the farm or by collection from the immediate environment.

If it is produced on the farm, it is commonly a by-product (see Table 4.5 for examples), straw and animal manure being the most important sources.

TABLE 4.5
FARM BY-PRODUCTS WHICH COULD BE USED FOR FUEL
(After Carruthers and Jones, 1983)

By-product	Fuel type
Dry crop residues, e.g. straw	Solid
Wet crop residues	Biogas
Animal wastes	Biogas
	Solid (Third World)
Horticultural wastes	Solid/biogas
Wood residues	Solid

Increasingly, however, it may be that fuel will have to be deliberately produced on the farm, by growing fuel crops, including trees, or by converting biomass to methane (by digestion). If it continues to be collected from surrounding areas, it will lead to deforestation, destruction of the vegetation and soil erosion, whilst occupying many hours of work in collection.

In developing countries, it can be grossly misleading to think about the improvement of food production without also thinking about how it is to be cooked. A systems approach ought to ensure that the wider view is, at least, always considered. (The same argument applies to water supplies.)

In developed countries, the cooking of food and the heating of houses is not based primarily on biomass but on electricity and gas, derived from fossil fuels.

Nevertheless, active consideration is being given to the use of biomass for fuel, on the farm and as a contribution to the national supply. Methane generation is being used but mainly from sewage works or town waste dumps, wood stoves are used to an extent that varies greatly from one country to another, and straw is burnt, usually close to where it is produced.

All these activities actually represent a move towards use of current rather than past biological processes.

The question that must now be asked is how should these biological production processes be described in order that they may be *improved?* Clearly, the answer hinges on what is meant by improvement.

One thing appears obvious: we are not interested in producing *more* of any of these products, except per unit of some resource (land, labour, capital, solar radiation, water, nitrogen . . .). So, as argued in the previous chapter, improvement must mean an increase in the efficiency with which *some* resource is used. It is here that the difficulty lies. We cannot usually have greater efficiency in the use of *all* resources at the same time, largely because one resource will probably be most efficiently used when all others are in plentiful supply and thus non-limiting. It is not inevitable that a resource in plentiful supply should be used inefficiently but some of it will be surplus and there is thus opportunity for loss and wastage.

On the other hand, the optimum use of all resources requires a weighting of individual resources such that one is regarded as worth more or less than another. This is most commonly done in monetary terms but there are risks in using current costs and prices since these can change quite rapidly. A range of costs and prices can be used but the results can only be related to a known or stated selection of them. One of the arguments in favour of calculations based on energy is that it is the ultimate driving force in biological systems and will probably be the major influence on future costs in agriculture.

Certainly, the primary role of plants in agricultural systems is the *fixation* (not merely the *use*) of the energy in solar radiation, for subsequent use, directly or indirectly, as a fuel for human activity. The form in which it is fixed, and the compounds made available as a result of plant growth, vary with crop species, as does the amount produced, but this is also influenced by numerous inputs, not all of them biological.

Once a process has been identified as specifically as this it is likely that the most useful description will take the form of a model that includes all the relevant interactions. Only in this way is it possible to determine what change in what part of the process would actually constitute an improvement or, if the model relates to only a small part of an agricultural system, a possible improvement. Much biological research in the general field of agriculture is really concerned with the establishment of biological or technical possibilities.

Similar considerations apply to animal production.

The Role of the Animal
Some of the original roles of animals have greatly diminished in importance in countries with developed agricultural and industrial systems but they are still very important in many parts of the world. These are:

(1) Transport (carrying burdens, including people).
(2) Traction (pulling vehicles and implements).
(3) Work, other than (1) and (2) (providing power for water pumping, threshing and grinding).
(4) Special functions (treading, hunting, warfare, guarding, herding, guiding the blind, tracking) although some of these are increasing (especially such modern purposes as drug detection).

The roles which are most central to agriculture are those concerned with production and they could be classified by products or by the species of animals used.

Numerically, the most important agricultural animals in the world are cattle, sheep, pigs, goats, buffalo, horses, asses and mules. Poultry are important in many countries and, in *particular* countries, the most important animal might be the llama, the reindeer or the camel.

The range of products includes meat, milk, eggs, fish, honey, leather, fibre, fur, silk, bone and faeces.

If we now ask why we keep a particular animal, it will probably be correct to answer that it is in order to produce a particular product—either because we want that product to consume or because we can sell it.

However, few animal products are now absolutely vital and they are rather costly to produce, so it is worth looking at the role of animal production somewhat differently, in terms of what special functions animals perform. The first function is nevertheless mainly concerned with production:

(1)   Output, including production and performance.

(2)   The collection of plant food (that is otherwise unavailable, impracticable or uneconomic to collect directly).

(3)   The conversion of plant material (especially cellulose) to more useful or more attractive forms.

(4)   The concentration of nutrients—this generally accompanies the conversion process and is often accompanied by the elimination of undesirable or toxic materials.

As with crop production, the identification of the important biological processes needs to be followed by a description of them in terms which allow an assessment of the effects of changes within them or to their inputs. This will generally lead to a modelling approach and, if the information available is adequate, can lead to the choice of the most desirable systems or to the improvement of existing systems. What is most desirable still has to be defined and the same problems occur, as with crop production, of balancing the efficiency of the use of one resource with that of others.

If the available information is not adequate, the information needed can be clearly identified: this has important implications, of course, for the planning of research programmes and the determination of research priorities.

The description of animal production systems must also identify the main biological processes on which they are based. These are activity (including feeding, digesting, excreting and resting), growth and development, reproduction and secretion. Senescence and death also occur but the latter is partially controlled within agricultural systems (or its timing is!). These processes can be further analysed in order to identify other processes (milk fat synthesis, cell division, ovulation) at a more detailed level of organisation.

BIOLOGICAL EFFICIENCY

Biological efficiency refers simply to the operation of these biological processes, expressed in terms of ratios of selected outputs and inputs. Neither of the latter *has* to be a biological material, although they often will be: a quantity of water pumped per hour may be a measure of the efficiency with which a biological process (applied animal power) is operating. Milk yield per cow is more obviously biological: so is silk output per unit weight of mulberry leaves, although this will sound less familiar to many people.

Clearly, the possible number of interesting and important ratios is very large indeed and cannot be dealt with comprehensively here. Two examples have therefore been chosen to illustrate the kinds of efficiency that are measured.

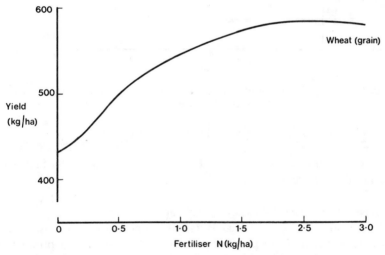

FIG. 4.2. The response of wheat to fertiliser nitrogen (after Norman and Coote, 1971).

(1) Fertiliser Use by Crops

Fertiliser has to be purchased by the farmer and, even if natural manure is used, work is required to apply it. These costs have to be justified by increased output, so the relationship between output of crop and input of fertiliser is very important. The three major fertilisers are based on nitrogen, potash and phosphate. To non-leguminous crops, the supply of nitrogen is usually the most important quantitatively, although such a statement is clearly nonsense where another plant nutrient is in limiting supply. The response curve to nitrogen supply is shown for wheat in Fig. 4.2. In most cases, applying fertiliser nitrogen to a legume merely reduces the legume's own ability to fix nitrogen in its root nodules and may not result in any greater production than before.

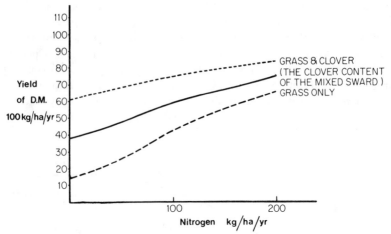

FIG. 4.3.   The response of grass and grass/clover swards to applied nitrogenous fertiliser (after Spedding, 1970).

The response of many grass pastures that contain various proportions of clover is therefore partly due to greater grass growth and partly to reduced nitrogen fixation by the clover (see Fig. 4.3): at very high levels of nitrogen application, the legumes actually die out.

(2) Feed Conversion Efficiency by Animals

The second example takes feed as its starting point since the cost of feed usually represents a high proportion of the total costs of animal production (Table 4.6). The use made of feed is therefore of great biological and economic significance.

TABLE 4.6
FEED COSTS AS A PROPORTION
(%) OF TOTAL COSTS

Production system	%
Pork	80
Beef	85
Milk	50
Eggs	75

(Source: Børgstrom, 1973).

If we measure the amount of feed (energy or protein) consumed by an animal and measure the energy and protein produced in its products, we find considerable differences between species (Table 4.7) but, of course, not all these species could live on the same feeds or in the same environments. So it may be more useful, at times, to have sheep producing protein with an efficiency of 4% from grass, which we cannot eat, than to have a hen producing with an efficiency of 20% from grain that we could eat (although it is worth noting that the comparable figure for a cow is about 24%).

However, individual animal efficiencies are only part of the study, since breeding populations always have to be maintained by someone somewhere.

In most cases, the efficiency of feed conversion is reduced when expressed

TABLE 4.7
EFFICIENCY OF PRODUCTION BY INDIVIDUAL ANIMALS
(Sources: Large, 1973; Spedding and Hoxey, 1975)

Animal product	Efficiency	
	Energy $\left(\dfrac{energy\ in\ product}{energy\ in\ feed}\right) \times 100$	Protein $\left(\dfrac{N\ in\ product}{N\ in\ feed}\right) \times 100$
Cow's milk	20	17–42
Rabbit meat	12·5–17·5	34
Beef	5·2–7·8	8
Lamb	11·0–14·6	16·4
Hen's eggs	10–11	16–29
Broilers	16	30
Pig meat	35	25–32

TABLE 4.8

FEED CONVERSION EFFICIENCY BY ANIMAL POPULATIONS
(calculated for breeding units of one female plus progeny and including
the relevant proportion of the feed intake of the male)
(Sources: Large, 1973; Spedding and Hoxey, 1975)

Animal product	Efficiency (ceiling values)	
	Energy $\left(\dfrac{energy\ in\ product}{energy\ in\ feed} \times 100\right)$	Protein $\left(\dfrac{N\ in\ product}{N\ in\ feed} \times 100\right)$
Cow's milk	12–16	40
Rabbit meat	8·0	23–40[a]
Beef (suckler)	3·2	9
Lamb	2·4–4·2[a]	6–14[a]
Hen's eggs	11–12	24
Broilers	14·6	25–26
Pig meat	23–27	17–22

[a] Depending upon prolificacy.

per unit of a breeding population (Table 4.8) since the feed required to support breeding females and the necessary proportion of males has to be included, and for the whole time, not merely when they are being productive. The minimum breeding unit must include the necessary males and females but these animals do not live for ever and so have to be regularly replaced. Any assessment of the efficiency with which feed is used by an animal population therefore has to take into account the feed used to produce replacement breeding stock, output in the form of old breeding stock being replaced and losses due to disease. Population efficiencies thus tend to differ somewhat from those of simple breeding units (Table 4.9).

However, as mentioned earlier, different species vary in their capacity to utilise different feeds and not all feeds can be grown on all soil types. Animals and crops also differ in the climatic conditions under which they can be kept or grown. The relative efficiency with which land is used may therefore be quite different from that calculated for feed use.

On arable land, for example, crops or grass could be grown and converted by animals. In some cases (barley, for example) the same animals, and in other cases (grass, for example), only a restricted range of animal species, can be used. Comparisons can be made in several ways, therefore, and the relative efficiency per unit of land expressed accordingly (Table 4.10).

However, it will be clear from the first part of this chapter and from the

TABLE 4.9

$$E = \frac{\text{N output in product(s)}}{\text{N input in feed}} \times 100 \text{ (per annum)}$$

	E % (approx.[a])
Rabbit	42
Dairy cow (for milk)	39
Broiler fowl	25
Hen (for eggs)	18
Sheep (including wool)	17
Sheep	14

[a] The values depend greatly on the levels of performance assumed for each species. In this calculation, the levels were as follows:

Rabbit	50 progeny per doe per year
Dairy cow	10 000 kg milk per cow per year
Broilers	120 eggs per parent hen
Hens (for eggs)	250 eggs per hen per year
Sheep	6 lambs per ewe per year

TABLE 4.10

ANNUAL CATTLE OUTPUT PER HA OF GRASS[a] OR BARLEY[b] based on feeding a mixed ration with different proportions of dried grass
(after Homb and Joshi, 1973)

Percent of dried grass[a] in the ration	Bull carcasses[c] produced (kg/ha)	Milk[d] (kg/ha)
21	450	
30		6 177
39	490	
45		6 765
55	530	
60		7 353
69	580	

[a] Grass yield assumed to be 10 000 kg DM/ha.
[b] Barley yields assumed to be 3 500 kg DM/ha.
[c] Based on bulls producing carcasses of 240 kg.
[d] Based on cows producing 6 000 kg of milk.

earlier chapters that the efficiency with which one resource is used usually influences the efficiency with which all the others are used, and some of the latter may be as important as the one first examined.

Feed is an important resource because it represents a high proportion of the total costs of animal production. Land is important because it is ultimately limiting but, at any one time, there are many parts of the world where additional land is available. Support energy is now recognised as of enormous importance because it is non-renewable, limited in quantity and likely to rise in price. If fresh supplies are discovered it only alters the time-scale and, because the cost of extraction tends to increase disproportionately (since new sources are generally in less accessible sites), may not greatly affect the rate of price increase. Since obvious alternatives are difficult to envisage on the necessary scale, it follows that the efficiency with which such a resource is used must be carefully considered.

It will also be obvious by now that support energy inputs have often been used to improve output per man, per unit of land and per unit of solar radiation received. Attempts to increase the efficiency with which support energy is used may therefore reduce all these other efficiencies, although

TABLE 4.11

EFFICIENCY OF USE OF SUPPORT ENERGY IN CROP AND
ANIMAL PRODUCTION

(Compiled by J. M. Walsingham: see Spedding, 1976a)

Product	Efficiency $= \dfrac{\text{Gross energy in product}\,^a}{\text{Support energy input}}$
Rice grain	3–34
Wheat grain	2·2–4·6
Maize grain	2·8–54
Potatoes	1·0–3·5
Milk	0·33–0·62
Eggs	c. 0·16
Lamb (plus wool)	c. 0·39
Beef	c. 0·18
Broiler hens	c. 0·11

^{a} Actual values vary with yield and with the method of calculation.

there are ways (such as reduction of wastage and losses) that may benefit all these ratios.

Some animal production systems use more support energy than others and, in general, animal production uses more and uses it less efficiently than crop production (Table 4.11). In the use of support energy, as with other resources, it is possible to work out where it is used in the production process and in what quantities (see Table 4.12 for example): it differs from most other resources, however, in also being involved in the use of most of them and in the sheer number of points at which it is a significant input.

TABLE 4.12

THE DISTRIBUTION OF SUPPORT ENERGY USE IN AGRICULTURAL SYSTEMS

(Table constructed by J. M. Walsingham to show % of total support energy cost attributable to each input)

Inputs	Production system			
	Beef	Milk	Potatoes	Barley
Fertiliser N	55	60	41	41
P	4	2	8	2
K	2	3	5	3
Machinery manufacture	8	< 1	10	27
Field operations	12	20	17	20
Herbicides	< 1		4	1
Grain drying	15			6
Electricity	2	15		
Seed	2		15	
Sources:	Bather (1975)		White (1975)	Austin (1975)

Although such 'energy accounting' is different from accounting in monetary terms, it is often defended on the grounds that it may represent future economic conditions rather better than would current costs and prices.

In any event, it is an interesting example of a form of analysis which is neither wholly monetary nor wholly biological, even when applied to primarily biological processes.

In a sense, it is a form of economic analysis, dealing with the efficiency of use of a scarce and costly resource for the benefit of Man.

THE MEASUREMENT OF EFFICIENCY

Efficiency can be measured in many different ways, not merely because there are many different ratios but because methods of measurement and expression also vary.

There are three major considerations which should be taken into account in both measuring and interpreting efficiency ratios. They are:

(1) The terms in which the numerator and denominator are expressed. (The actual figure for a milk production efficiency ratio will be different if it is expressed in litres per year or lb/year.)

(2) The period over which a ratio is measured. (Reproductive efficiencies of elephants and mice are better compared over different time periods: efficiency of the growth of grass will be different over one month and one year and for different months.)

(3) The environment in which the efficiency is measured. (The feed conversion efficiency of a dairy cow is likely to be greater than that of a sea-cow—but not notably so in a marine environment.)

THE ENVIRONMENT

All biological processes take place within variable environments and improvement has to be related to the cost and difficulty of imposing whatever greater control of the environment may be required to achieve the improvement.

It is clearly of little use devising new ways of operating plant and animal production systems if these are going to cost too much or require impracticable modifications to the environment. On the other hand, how can we possibly know that they will cost too much if we cannot even visualise the innovations because they have not yet been devised?

This polarisation of views about applied research has seriously hampered agricultural investigation, with some arguing in favour of pure science and unfettered exploration (from which useful ideas can subsequently be selected), and some arguing that the practical context is the only one in which useful solutions can be found to real problems.

Having considered briefly the biological processes underlying agriculture, we are now in a position to discuss the contribution that science and the scientific method can or should make to the development of agricultural systems (see Chapter 6).

First, however, it is worth considering the efficiency of agricultural systems in economic terms.

5

Economic Efficiency in Agriculture

Since agricultural systems have to support those who operate them, they are bound to be subject to certain basic economic principles. The latter apply to farm businesses just as to any other business activity, although it is often true that the objectives of farming may be rather different. If a large tract of land is wholly owned, it can be used for very low output enterprises (or not at all), such as grouse or tiger hunting. It may be said that the output of pleasure is substantial but clearly it need not be so. In such cases, of course, the purchase of the land was not part of an agricultural operation at all and it is necessary to distinguish between land ownership and agriculture: they interact but they are clearly not the same.

It is often thought that the basic principles of economics give rise to the most startling differences between biological and economic efficiency, so it is worth considering how these principles apply to agriculture. Two of the most important will be taken as examples: (1) the law of supply and demand and (2) the law of diminishing returns.

(1) THE LAW OF SUPPLY AND DEMAND

In biological terms, the efficiency of an agricultural activity is not influenced by whether anyone wants the product or is prepared to pay for it.

Economic efficiency, however, may easily be dominated by such factors. If no-one buys the product, there is no income and farming eventually ceases. Indeed, if the total supply exceeds the total demand, the price can fall to a level that does not adequately cover all the costs of production.

The biologist looks at the hungry people of the world and is convinced that more food must be produced, but if the hungry are too poor to

73

purchase the food no benefit necessarily follows. Nor can food be given away free, except in limited circumstances, without risk to the livelihood of the producer. There are, of course, problems of food being produced at a distance from where it is needed or at the wrong time and there are thus real problems of storage, preservation and distribution. Even so, the major discrepancies between what makes biological sense and what makes economic sense may often be social in origin.

The biologist will often react by arguing that it must nevertheless be right to learn *how* to produce more food, even if the world is not always able to use this information immediately. It might be better, however, to rephrase the problem and state it as how to produce more food for those in need, within current or feasible economic and social frameworks.

These will always include some need to relate supply and demand of each product and to recognise that demand does not simply reflect *need* but reflects what the consumer wants and what he can afford to pay. Furthermore, these last two are related in ways that differ for different people. If each of us obtains more money, the range of goods that we want will probably increase, but exactly *what* we want and how much we want it and whether all our additional money will be spent on additional wants are all things that vary from one individual to another.

Such attitudes apply to all manner of goods: there is no simple division between essentials and luxuries. Food is certainly essential but no particular item can be so described unless the range of foods available is very restricted—which may often be the case, of course, for physical or economic reasons. In most circumstances, if money is available, there are many ways in which the essential nutrients can be supplied, including many different sources and many different ways of processing, preparing and cooking foods. However, most people also enjoy eating and, if they can afford it, may eat much more than they require. The same sort of arguments can be applied, for example, to clothing and housing.

So economic demand for a particular product is not easy to predict because it also depends on what other products are available, not necessarily as strict alternatives but competing for our choice. This notion of competition runs through much of the economic activity of agriculture. Farmers are, in general, small businessmen, competing with each other. It is true that there are some very large farm businesses and even small farms may now involve enormous capital investment. Nevertheless, most of the farmers in the world exist in family units and if they overproduce the price falls. This often leads to cycles of glut and shortage, particularly well known in pig production (see Fig. 5.1). It has also led to farmers banding together,

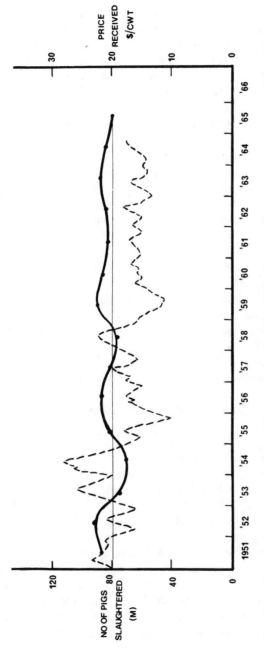

FIG. 5.1.   The 'pig cycle'; —— number of pigs slaughtered in USA, 1951–65; – – – approximate price received ($/cwt) (adapted from Davidson and Coey, 1966).

in co-operatives, or setting up marketing organisations to work towards orderly marketing arrangements. However, competition is not merely a matter of one pig farmer competing with another: if the price of bacon is high, the consumer may decide to buy beef or lamb instead. Indeed, he does not have to eat meat at all and could live on crop products.

Unfortunately, all the farmers in one area may well be producing the same range of products and yields may all be influenced by the same weather. So they may all have a good year for apples, or a bad one for potatoes, and violent swings in price may be unavoidable. This can happen on an enormous international scale and a purchase of wheat by Russia from America can have far reaching consequences.

If we wish to feed people adequately, it is essential to produce enough, even in the bad years, and this means overproduction in the good years, since we cannot predict which sort of year is coming, climatically. Storage sounds the obvious answer but it is often very expensive and it may be cheaper to move food about the world, from areas of surplus to those of shortage. All this implies some planning and a reduction in the degree of competition to which food production is characteristically exposed, largely because of its dependence on the weather, although unpredictable outbreaks of disease also have similar effects.

However, the farmer does not necessarily aim to produce maximum yields per unit of land, because he has to pay for all the resources used and yield may not rise in simple relation to input of resources.

(2)   THE LAW OF DIMINISHING RETURNS

This law states that the yield response to inputs eventually diminishes, so that, after a time, each additional quantity of input results in less additional yield than the previous one. It is a quite general proposition and obvious examples are easy to find (see Figs. 5.2 and 4.2 for illustrations of the response of sugar beet and wheat to nitrogenous fertiliser), but it nevertheless has to be interpreted with care.

In the case of grass, for instance, the response (in yield of dry matter) to applied nitrogenous fertiliser is, for all practical purposes, linear, although there is, of course, an upper limit beyond which there is no response at all. If we consider the feeding of warm-blooded animals, however, there is no production response at all (whether of eggs, milk or meat) to inputs of feed until the maintenance needs of the animal have been met: after this point, the response tends to be linear for a substantial period but, again, there is a

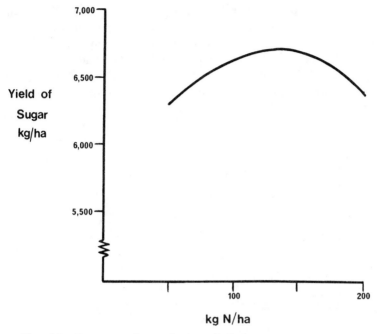

FIG. 5.2. Response of sugar beet to applied nitrogenous fertiliser.

limit to yield potential. An importance difference with animal feeding is that there is a limit to the quantity of feed that can be consumed, whereas there is no real limit to the amount of fertiliser that can be *applied* to a crop. (There is, of course, a comparable limit to nutrient uptake by crops.)

The law of diminishing returns is thus rather a crude statement of what is likely to happen and it is usually better to uncover the biological relationships that are being modified.

Plants and animals require a range of inputs, including nutrients, gases such as oxygen and carbon dioxide, water and heat, and these are rarely all present to excess. Naturally, if everything is available in plentiful supply except one element, supplying this one element will usually lead to a response. The plant or animal grows and, if the situation remains substantially the same, further supplies of the limiting element will lead to further growth. Eventually, of course, some other element becomes limiting and the response to supplies of the first one diminishes. In some cases, supplies in excess of need may be harmful: indeed, if the excess is great enough (consider water, for example) this is inevitable.

It may well be the case that, for some inputs, the response will be linear up to some optimum value at which it immediately becomes negative. Response to ambient temperature, reflecting the supply of heat, may be in this category (see Fig. 5.3 for examples relating to fish and cattle), although there may be a brief period before the optimum is reached, during which diminishing returns can be observed. (For a discussion of the analysis of response in crop and livestock production, see Dillon, 1977).

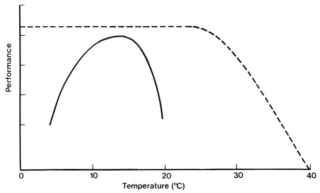

FIG. 5.3.   Response of fish and cattle to temperature; —— growth of temperate fish; – – – – milk production in cattle (after Weatherley, 1972 and Hafez, 1968).

In practical agriculture, situations are usually complex and not a great deal may be known about the status of various nutrients in the soil or even in the feed of animals. The law of diminishing returns supplies a rough rule to remind the farmer that, just because he gets a response to one level of an input, it does not follow that he will get a similar response if he applies some more of the same thing.

Since economic efficiency depends upon the relationships between the value of the outputs and the cost of the inputs, it is obviously important to understand what response may be expected from each additional quantity of input. This is known as marginal analysis but it has to be based on the underlying biological responses and it is often difficult to put a monetary value on these.

THE VALUE OF BIOLOGICAL RESPONSES

The response of a whole organism may well be represented by a quantity of product, such as grain, milk or eggs, or by an increase in number of fruits or in animal weight. The latter can be visualised as representing intermediate products, on the way to the products that are eventually sold. As such, it may be possible to put a monetary value on them. This can readily be done for whole products and can certainly be done for virtually all inputs.

Furthermore, plants and animals are also sold from one farmer to another, so it is possible to value seedlings, seeds, pregnant cows and weaned lambs, even though they do not represent products as sold to the ultimate consumer.

In all such cases, a monetary value not only reflects the current biological value to the farmer, relative to the costs and values of other inputs and outputs, but is probably the only way in which such a value can be expressed. This simply reflects the function of money.

There are, however, biological processes that are extremely difficult to evaluate in this way and yet are vital to productivity.

This is clearly so for many processes that are internal to plants or animals—such as the daily rate of nitrogen fixation in legumes or the digestive efficiency of a cow—but also applies to other organisms, such as earthworms, dung-beetles and algae on the soil surface, that never enter into any form of trade. In these cases, no economic value can currently be established, although it is possible to imagine circumstances in which this could be done. Part of the problem is to relate these processes to the business of farming at all and herein lies the clue to the most important link between biological and economic efficiency. It depends upon establishing the relevance and importance of a component process to the functioning of the whole system—the essential proposition within the systems approach described in Chapter 2.

This is not to say that biological and economic efficiency are really synonomous when referring to the same system. For example, it may be biologically more efficient to produce white eggs, in the sense that more human food is produced per hen, per unit of feed, per unit of monetary cost or per man. It may still be more economic to produce brown eggs if, in spite of there being no biological or nutritional advantage, people are prepared to pay a higher price for them. Nor is this in any way ridiculous when compared with the enormous variety of ways in which we are prepared to pay more for foods we prefer, even to the point where people who are slimming may actually pay more for food containing *less* energy.

What *is* being suggested is that the biological efficiency of a component process cannot be measured independently of the total system if it is expected to relate to it.

Reproduction in sheep will serve as a convenient example. Figure 5.4

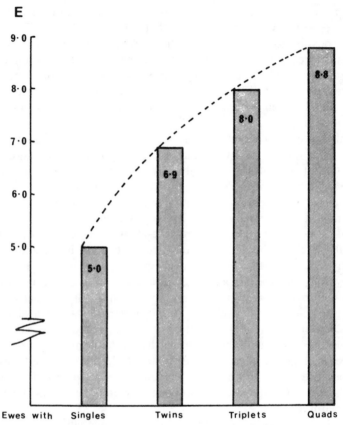

Fig. 5.4.   The effect of reproductive rate on the efficiency of feed conversion (E) by sheep:

$$E = \frac{\text{output of carcass (kg)}}{\text{digestible organic matter consumed (100 kg)}}$$

shows that feed conversion efficiency is increased by increases in reproductive rate but it cannot say anything about economic efficiency, even of the same system, because it does not take account of any additional labour or medicines that may be required, or of any increase in the cost of feed due to the need to purchase milk substitute to feed the extra lambs produced at high reproductive rates.

Figure 5.5 shows the increase in the number of lambs born in response to injections of pregnant mare serum (PMS). Now, here, even taking account

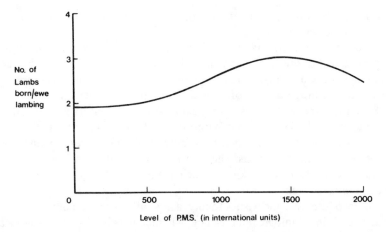

FIG. 5.5. The influence of pregnant mare serum (PMS) on prolificacy in sheep.
(Source: Newton and Betts, 1966, 1968).

of the cost of PMS, we cannot deduce anything about the biological
efficiency of the whole system because no information is given about, for
example, mortality in relation to reproductive rate (see Table 5.1).
Wherever it is possible to relate a change in a component process to a
change in the whole system, it is possible to calculate a relevant biological
efficiency: this can be expressed in economic terms, for the whole system,
only if *all* the costs and prices are included.

The idea of maximising biological efficiency has no more relevance within

TABLE 5.1

LITTER SIZE AND EARLY MORTALITY IN SHEEP TREATED WITH
PREGNANT MARE SERUM (PMS)

(From Spedding (1970); data compiled by J. E. Newton)

Litter size	Percent mortality	Number of lambs alive at 1 day per ewe lambing
1	6	0·94
2	8	1·84
3	10	2·69
4	27	2·94
5	13	4·33
6	67	2·00

economic systems than any other proposition that relates to only some of the resources employed.

As a matter of fact, it is extremely rare for anyone to want to maximise anything, without regard to anything else. Farmers are often thought of as wishing to maximise profit. This is plainly not so in absolute terms and we have said nothing about profit per unit of what (year, hectare, cow, etc.).

Even if we arrive at a satisfactory expression of monetary return, hardly anyone will wish to *maximise* it, without regard to the effect this has on life-style, satisfaction of the farmer, his family or his employees. So, although most people want to increase the returns to their business, this is hedged about by constraints of all kinds—such as the wish to take holidays or do other things as well.

But insofar as anyone wishes to increase profit, it does not always follow that increasing biological efficiency will achieve it. So the biological improvements that scientific research may lead to, may not actually benefit the business. It is this difficulty of establishing the relevance of biological 'improvements' and the fact that agriculture includes many non-science features that make for some confusion as to just what science can contribute to agricultural development.

6

The Contribution of Science

Since agricultural systems contain non-scientific components and unquantifiable relationships, it often appears that science can only have a limited role in their improvement. However, orthodox science subjects also contain relationships that can only be quantified in a statistical sense, so that prediction is concerned with an estimate of probability.

The important questions, really, are whether the knowledge required for the improvement of agricultural systems is scientific or not and whether such knowledge can be applied within a practical context.

THE KNOWLEDGE REQUIRED

In practical farming, which is, of course, only one aspect of agriculture, important questions are often unanswerable with any degree of confidence or precision.

For example, will it be more profitable to plant beans or barley on a particular field in a particular year? No answer can be given unless a great deal of additional information is provided, including such data as future costs and prices, future weather, disease incidence, demand for the product and even the occurrence of accidents (to the farmer or to his equipment). Furthermore, such information has to be predicted well before sowing, or even before decisions are taken about cultivation or the ordering of seed and fertiliser.

But no-one and no method can answer such questions by the use of reason or logic. Science can only answer questions that are posed in a particular way and this must include a statement of all the values to be attached to all the relevant variables.

Clearly, agriculturally important questions can be framed in this way and are no different from those in any other field in which values have to be postulated in the question, whether they can be predicted for some future date or not. However, it is not obvious over what range of subjects these scientific questions can be posed. For example, a biological or chemical question would be assumed by everyone to be acceptable but an economic question might easily be considered inappropriate.

It is difficult to defend this kind of distinction. There seems to be no fundamental difference between a 'biological' question concerned with the use an animal makes of its feed and an 'economic' question that relates a quantity of output to a quantity of input, however these quantities are expressed. Monetary expressions are no more arbitrary, necessarily, as a means of weighting things according to some scale of values, than expressions of quantities on a dry matter (or protein or energy) basis because we consider that this expresses the 'value' of the product more accurately than, say, the weight of wet roots or the liveweight of an animal.

If the questions that science can deal with cannot be distinguished in this way, then how *can* they be distinguished? The only satisfactory answer seems to be the rather circular one that science can deal with questions that are susceptible to the scientific method: this has chiefly to do with how scientific 'proof' is established (discussed later in this chapter). Experimentation is one obvious way of establishing the truth of (or of failing to disprove) a proposition but it will be recalled that this is really too limiting to be a necessary condition. In general, the 'truth' of a proposition may be regarded as probable if it is repeatedly observed under reasonably controlled conditions. If this is associated with a probability that it will occur in given conditions, it is clearly valuable information and can be used to practical benefit, independent of whether it relates to animals, plants, weather, men or money.

The acquisition of knowledge that relates to recognisably agricultural systems and is obtainable by scientific methods can legitimately be termed agricultural science. Knowledge that relates to the agricultural usage of soils, plants and animals may similarly be regarded as within the province of agricultural science. Knowledge about agricultural plants and animals, however, may be biological rather than relating to agricultural science.

Thus scientific knowledge of various kinds may be useful agriculturally, even though agricultural science may be distinguished as a separate activity. Biological knowledge, for example, may be used by an agricultural scientist in the course of his own work: furthermore, he cannot necessarily know in advance which bits he will need.

For example, it might be thought that the manner in which an animal deposits its faeces is of no great agricultural significance. A moment's thought shows how wrong this is. Faeces may be the vehicle of disease and parasites, they may represent a disposal problem or a source of pollution, or they may foul the animal's own environment, including its feed and water supply.

Purely biological studies might reveal that, for example, the donkey tends to deposit its dung on top of existing dung pats. This might lead to speculation that other animals might exhibit characteristic, and perhaps quite different, excretary behaviour. And so they do. Cattle and sheep tend to defecate anywhere and everywhere, although night camps may lead to accumulations. Horses always use a particular part of the field, pigs can be trained to dung in specially provided passages and hens tend to defecate just before laying. Much of this information is not only useful but is already made use of in the management of both housed and grazing livestock.

There are therefore three main kinds of scientific knowledge that are relevant to agriculture.

First, there are the fruits of agricultural science, largely dealing with agricultural systems or sub-systems.

Secondly, there are the results of other scientific activities (biological, physical, etc.), carried out on, or with, agricultural components or component processes but with no agricultural purposes or applications directly in mind.

Thirdly, there are the results of other scientific activities using non-agricultural components and processes: these may give rise to the second kind of knowledge, either indirectly or by new components and processes being introduced into agriculture.

METHODS OF OBTAINING KNOWLEDGE

It has been implied that if science is to contribute to agriculture it must be by the acquisition of relevant scientific knowledge by scientific methods.

The practical man usually believes that there is here a serious risk of ignoring important information that has been gleaned over many years of practical experience and observation, either in the sense that it is not used or that it is ignored until some science-based proposal falls flat because the existence of information based on experience was simply not recognised.

The first answer to this is that scientific knowledge is not the only kind that is valuable or used in most walks of life. The scientist does not generally employ scientific knowledge when crossing the road and people without scientific knowledge are likely to be at least as successful in doing so. We all use experience of many kinds and exercise judgement supported by the available, inadequate data, daily and without ever supposing that we should not do so because our knowledge is not scientific. Furthermore, even if we are considering an area in which scientific knowledge can be used, there is no reason to exclude the use of all other available knowledge.

In some cases, the difference will be in the degree of probability that something will happen, but quite often we are only concerned to minimise a risk. Thus, if we do not attempt to cross the road when a 'bus is within sight, it may have little to do with an accurate knowledge of speeds and distances. It may simply be that we can estimate that a 'bus could *not* knock us down if it is *not* within sight when we start. Farmers exercise this kind of judgement in relation to operations such as cultivation, harvesting, hay-making and the weather.

The second answer concerns the process of the application of scientific knowledge in practice. Sometimes it is legitimate simply to 'try it out in practice' over a wide range of conditions and periods of time and judge empirically whether it works sufficiently well or often to be worth adopting more widely. This approach assumes that there may be factors that have not been adequately taken into account up to that point and that the risk of going ahead directly is small. In other cases, a different approach is desirable—of gradual application, with monitoring of progress and continual modification to suit the conditions into which the innovation is being introduced. This is one way of looking at 'development' (in the research and development sense) and is best done by both those familiar with the science and those familiar with the relevant practice.

After all, the scientist will not claim as knowledge that something will happen in an environment he has not investigated just because it happened in the context he did investigate. He may have varied some of the factors of known importance and the practical context may be a relatively well-controlled one (such as a battery hen house), in which case he may feel that the same result is likely or the difference can be predicted. But if the practical environment is variable and some of its properties unknown, the only way to establish the validity of the scientific knowledge for that environment is to test it as a hypothesis. In short, it is not *known* for that environment until such tests have been completed satisfactorily.

In addition, it should be noted that experimentation involves observation and is sterile without it, whereas observation without experimentation is often difficult to interpret but is not useless. Consider the case of large predators and imagine an agricultural animal enterprise subject to the depredations of lions. Experience and observation could readily establish the cause of the problem and the probable solution: it is difficult to imagine an experiment that would improve upon this situation. Even so, it is quite possible that cheaper, or even more effective or more pleasant control methods could be devised than those which immediately spring to mind (high fences, elimination of the lions, feeding them so that they became satiated, or breeding very large, fierce guard dogs).

The fact is that the knowledge needed varies in precision as well as in kind: the depredations of a weevil on crop yield might prove very difficult to establish quantitatively without controlled experimentation and the precision required in the answer might be much greater. Since proof by experiment does not always seem to be required, it ought to be worth exploring just how much can be derived from practical experience and what ways are open to us to avoid faulty interpretation. In the case of the lions, the argument would be that 'seeing was believing' and it could not have been hyenas, for example, because the lions were observed killing stock (and these could be counted). It is as well to remember, therefore, that sometimes predators only remove the dying, the old and the diseased, that might have died even if the predators were excluded. So there are still interpretations to be put upon the observed facts and controlled experimentation is usually designed to restrict the possible number of interpretations and to indicate one of them with a high probability of being correct.

When it comes to quantifying this probability, it is difficult to do so without experiments designed for the purpose.

One of the things that can *only* be learned from observations of the relevant situation is the kind of familiarity with the material to be studied that must accompany the first steps in dealing with something new. Somehow a rough idea of the system encountered, its components and the way in which it works has to be built up in the mind before sensible experiments can be planned at all.

This must, therefore, be a part of science and a part of the scientific method: since it could equally well apply to non-scientific systems (including people and money), the scientific method probably ought to be applied completely if it is to qualify as scientific.

It is interesting to reflect on the enormous difficulty of ever carrying out controlled experiments on an individual, for everything that is done to it may have the effect of altering it, so it is not the same when the next thing is done to it. Indeed, it often seems that science has little to do with individuals (whether fields, animals or people) but this probably reflects the very low probability attached to propositions concerning individuals that tend to uniqueness: the greater the number of individuals, the greater the probability that can be attached to propositions about them. This emphasises yet again that scientific statements predict within stated ranges and do not represent the absolute certainties that science is popularly thought to deal in.

Most scientific methods of acquiring knowledge, therefore, are concerned with quantifying relationships observed under relatively controlled conditions.

THE CONTRIBUTION OF AGRICULTURAL SCIENCE

Clearly, if scientific methods can be directly applied to the study of agricultural systems or sub-systems, the knowledge gained should be of direct benefit. This pre-supposes sensible questions, of course, and it is not an easy matter to frame questions that are sensible, unambiguous, relevant and specific. Questions that are concerned with biology will not necessarily produce answers that will improve the profitability of farming.

Quite commonly, however, the most important questions from a practical point of view cannot be answered without more detailed information. It is at this point that relevance often gets lost and it seems necessary to build models of the systems to be influenced in order to establish what the relevant detailed questions are. Very often they will be designed to elaborate or quantify the models so that they can better serve the purpose for which they were constructed (often to answer the initial practical question).

It is obvious, however, that questions about existing systems (or variations on them) are not very likely to lead to revolutionary changes or radical innovation. It is hard to see how thinking about the practical problems of milk production could easily lead to the design of a system of fish farming, for example. A much more speculative approach is required for real originality and innovation. It can occur when agricultural scientists contemplate the results of pure biological research or when biologists contemplate agricultural applications of their findings.

Original thinking is almost impossible to plan, of course, and it would be hard to generalise even about the conditions under which it thrives. Amongst the latter, one might suppose that it would be profitable to have scientists thinking freely about the biology of agriculture and about animals and plants in this context (including those already used, but not confined to these). The context and the purposes of agriculture would probably need to be made clear by others, because the problems of agriculture have to be related to its purposes and these, as we have seen, are multitudinous.

So it would seem that, broadly speaking, agricultural science must define the problems and describe the systems (both in terms of specific, existing ones and in fundamental terms related to the objects of agricultural activity): it may also contribute to their solution, directly or indirectly, by using other scientific findings.

Biological sciences may, therefore, contribute to these solutions but, in addition, may give rise to quite new systems and processes. A balance of these activities would seem preferable to a predominance of only one.

At this point, it may be useful to consider in greater detail what constitutes science and the scientific method.

SCIENCE AND THE SCIENTIFIC METHOD

Science is usually thought of as a body of knowledge, relating to certain subjects in particular and to knowledge of a peculiarly reliable kind. Thus we talk of scientific knowledge, tending to mean 'known for certain', and of scientific subjects, meaning those with a basis of provable facts, usually expressed quantitatively. No-one argues that this is the only kind of knowledge, or the only knowledge worth having, but its distinguishing features are highly prized and jealously guarded.

Closer examination reveals that the 'certainties' are not quite what they may have seemed and that scientists may regard them as anything but the final truth. Indeed, scientists probably doubt whether anyone possesses the final truth about anything, except in matters of logic (2 + 2 *must* equal 4, for example), or could ever know whether they did or not. I say 'probably' because generalisations about groups of people are rather rash: 'scientists' are no more alike or uniform in their views than are farmers, students or readers.

A scientific proposition or 'fact' is characterised, not by being correct but by being testable or provable. If a proposition cannot be tested it cannot be called scientific. If it *can* be tested, it generally is, unless this demands immensely expensive equipment (as with some physics) or impracticable conditions (as could occur in astronomy) or intolerable activities (such as experimentation on human beings) or impracticable periods of time (as in some evolutionary propositions). It is generally regarded as important that tests can be made by many different people and it is accepted that only one 'disproof' of a proposition renders it false. Until proved false a proposition may continue to be regarded as true. Scientific facts are therefore those propositions that have successfully resisted all attempts to prove them false: but a new attempt could succeed at any time and the 'facts' would then undergo revision to restore the position, that they could not be shown (or have not been shown) to be false.

Scientists thus live with a healthy scepticism about their knowledge and recognise that it is almost certain to change with time. In fact, they are mostly engaged in trying to change it.

It may seem curious, therefore, that the common view of scientific knowledge is so different from that held by the scientist. The absolute certainty often read into science simply does not exist in the minds of its practitioners. What does exist, however, is the rigorous insistence that a 'fact' is not scientifically acceptable if it has not been exposed to this procedure of formulation in a testable proposition, followed by systematic attempts to falsify it.

A proposition that is testable but has not yet been sufficiently tested is called an 'hypothesis'. After a good deal of testing it may be termed a 'theory' and after a very great deal of testing, provided that it is of sufficient importance, it may be dignified by the term 'law'.

It is this insistence that a statement cannot be regarded as embodying any scientific facts or knowledge unless it has been through this scientific process that tends to imply undue certainty in those propositions that *are* accepted.

Since scientific truth and knowledge depend entirely on this testing process and since one clear falsification is deemed sufficient, it is not surprising that the testing procedures are themselves the subject of detailed and meticulous scrutiny. They usually involve both logic and experimentation or observation and they are subject to rules of procedure, carefully agreed conventions and terminology and a recognition that others must be able to repeat any test that is carried out. The whole enterprise is based on immense trust in the honesty of other scientists, combined with this capacity to check by repetition. The importance of this latter capacity cannot be over-emphasised. Facts are not acceptable if the observation leading to them cannot be repeated, given adequate conditions. Sometimes, of course, the conditions are not repeated and cannot be controlled. In general, however, if it cannot be repeated, it probably is not so. This casts a special light, incidentally, on talk of 'duplication', as if that was always undesirable.

Science, then, strongly implies not only the body of scientific knowledge but also the way in which this is acquired, since this is what renders it scientific. The scientific method is therefore a major characteristic of science and scientists and deserves very careful examination. Before discussing it, however, there are some other aspects of science that should be dealt with, including the way it is organised into subjects.

Subjects

So many areas of knowledge are referred to as subjects that it would be easy to regard the word as too imprecise to be meaningful. Other examples of these 'umbrella' words will be considered in the next chapter, because it turns out that they are quite useful. My own view is that a subject can best be defined as 'a focal point of interest, with that area of knowledge that immediately relates to it'. This was illustrated in Fig. 1.2 for agriculture as a whole and is illustrated by Fig. 6.1 for a constituent part of it.

One of the most obvious consequences of this way of defining a subject is that there is no limit to the number of such subjects. Furthermore, they may overlap with each other to varying degrees, but this accords with common sense. After all, if two people are interested primarily in the birds and the insects, respectively, in a garden, each one's subject is almost bound to

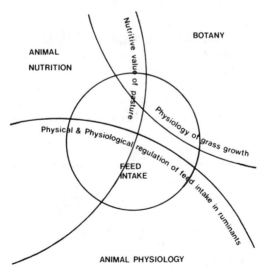

FIG. 6.1.   The delineation of subject matter within agriculture: illustrated for 'feed intake of the grazing cow' (as a subject).

include some features of the other's. If the birds eat insects, a study of the birds without the insects will be incomplete and, similarly, studies of the insects that never consider the birds would be deficient in important respects.

Another important feature is the frequency with which subjects will be inter- (or multi-) disciplinary. It is interesting to wonder why we should need to distinguish subjects and disciplines. Zoology, for example, is both a subject and a discipline, but the study of beetles may not be. The study of their anatomy would be, the study of their biology might not be (because it might include plants that they live on) and a study of their economic importance would certainly be multi-disciplinary.

What, then, is a discipline? It is clearly a branch of knowledge, or a subject, that is concerned with sufficiently similar constituents. It is the study of a collection of things, so similar that they can be grouped together as a class or category. They can usually be further classified or put into categories and this is often a major and central part of disciplinary activity.

Thus zoology is the study of all those organisms that can be called animals. Further classification may divide them into vertebrates and invertebrates, according to whether they have backbones or not, and into groups such as 'birds', not on the basis of where or how they live or what they do, for example (all things that fly are not classified together), but according to their anatomical structure and evolutionary relationships.

Chemistry, similarly, classifies its constituent chemicals according to its own criteria. So disciplines are based on classified similarities and differences but they depend upon the establishment of relationships of some permanent kind. The things that occur within one discipline can always be systematically classified and exhibit these permanent relationships. A beetle therefore belongs in zoology, whatever else it does, because it is a part of a collection of things with which zoology deals.

The study of a discipline has to recognise the rules that govern these relationships. Many of them will be important laws and it is the permanence of these laws that gives a discipline its strength and usefulness.

The classification of animals, for example, may seem rather a dull topic but just consider how helpful it is. Simply using the word 'animals' has already been able to tell you what we are discussing, in a quite extraordinary manner. You have discarded plants, crockery, soil, planets and cars, and focused on the animal kingdom without worrying about *which* animal, indeed without knowing all of them or perhaps much about any of them. But you know that they *all* have certain characteristics and that any statement that really applies to all of them should be immensely useful. If we go on and specify, for example, a mammal, you immediately know that the animal in question is warm blooded, gives birth to live young and feeds them on milk. These things are always so and the framework of such knowledge does, in fact, display a certain *discipline*, in the sense of laid down rules that cannot be broken. Now, anyone studying a multi-disciplinary subject is unlikely to be familiar with all the rules of all the disciplines involved. He may need to combine with someone trained in another discipline in order to deal effectively with the mixture of disciplines confronting him. It is thus of value to have established disciplines that serve as permanent reservoirs of a particular kind of knowledge about particular things, of special skills, laws, attitudes of mind and of a particular kind of people. The danger of such permanence is, of course, that of getting into a rut, of mistaking rigidities for useful frameworks (as one might come to view scaffolding as more important than the building it supports or encloses).

Disciplines, like knowledge itself, must be able to grow and evolve (see later section). It may also be that the present disciplines are not the only possible ones or even the best. Certainly, it should be possible to establish new disciplines, presumably when a subject becomes sufficiently large and important. However, size and importance by themselves may not be enough and the crucial criteria may relate to whether a subject has acquired its own unique set of principles and a coherent framework relating them to each other.

The Scientific Method

This term has been used for a long time to describe the characteristic way in which scientists acquire knowledge. In thinking about it, the meaning that scientists attach to knowledge is very important and there are some common misunderstandings about it that need to be sorted out first. For example, many people imagine that each subject consists of a 'core' of knowledge that can be regarded as known for certain and that this may be envisaged as surrounded by a peripheral area of less certain ideas still in need of confirmation. They think of research as being concerned with these peripheral areas, at the 'frontiers' of the subject, and that the aim of research is to extend these boundaries by the addition of new knowledge.

Now this might be so if research consisted simply of the addition of new information to some great existing heap—and there is sometimes a danger of this happening. But heaps of information do not constitute knowledge *about anything very important*. They represent isolated bits of knowledge and these are often important, but they relate to rather unimportant parts of the natural world. This problem is easily visualised in terms of the construction of a building. All the bricks, mortar, pipes, wires and planks needed to construct a building could be placed in a large heap and a great deal could be known about the weight, size, strength and composition of each unit, but there is no way that this would represent a building. The latter depends upon relating all these units to one another, and this depends upon further knowledge about the possible relationships between them and how they interact with each other.

This, like all analogies, has its limitations. Incidentally, some people regard analogies as dangerous: but so is a knife, or any other useful tool. Everything depends upon the educated and responsible use of tools. Analogies are intended to provide an insight that is difficult to acquire by direct study. So it is possible to help someone who has never seen a kangaroo to understand how it jumps by saying that it is a little like a frog. (This example is chosen partly because it is a *bad* analogy, but it does point to what a *good* analogy should do.)

The main limitation on the building analogy is that it involves too great a permanence in the structure once built. The creation of knowledge is continuous, or at least it never ceases, so the building is not only never finished, even its main elements change, but the addition of a new fact of central importance can change the very shape and structure of the whole building. Knowledge may be said to grow, therefore, in the same way as does an organism (e.g. like a man or woman), by the incorporation of new units (of food, material, information) into the existing body, involving the elaboration of this new material and its transformation into different forms in such a way that the body grows and develops into something that may be quite different from the original. (Note that we are embarked upon another analogy, the weakness of which is the implication of a pre-determined pattern of growth and development.)

It is not that knowledge cannot be gained about the periphery of a subject, but that this tends to be knowledge about relatively trivial (not central) issues. The most important scientific activity actually bears on the centre of a subject and may reshape the entire structure from within.

All this has a considerable bearing on the way in which scientists acquire knowledge and the role of particular activities, such as experimentation. The first thing to note is the magnitude of the intellectual activity that is required to comprehend whole spheres of knowledge sufficiently to reshape them. The second is that the *idea* of doing so (never mind how or in what direction) simply does not occur to everyone. Indeed it occurs to very few and it is only the intellectual giants who can stride, fearlessly and with a unique combination of humility and confidence, across such large areas of human knowledge. This intellectual activity I would put at the top of my list of activities characteristic of the scientific way of life. It is not, of course, confined to scientists but a list confined to the *unique* properties of scientific activity describes the latter very inadequately—as well try to describe a plant by the peculiar hairs on its seed that may distinguish it from all other plants or imagine that you know about chicken pox because you can diagnose it by the characteristic pimples it produces.

This discussion has seemed necessary in order to counterbalance the frequently encountered view that the scientific method is preoccupied with such things as experimentation. The latter is concerned with *how* new information is gained, in a rather practical sense, but the process of assimilating such new information and creating knowledge requires a great deal more than experimentation.

Experimentation

This is certainly a characteristic feature of the scientific method but is better preceded and followed by much thought.

A great deal of discussion is often centred on the choice of experimental method when, in fact, this should largely follow from sufficiently well thought-out objectives. However, in case this sounds too simple or too easy a solution, it is as well to admit to the extreme difficulty that often surrounds this first task of stating the objectives of an experiment.

Consider an everyday example. We wish to know how to grow more food. Superficially, this sounds a very clear and desirable objective, so why not do an experiment to find out? The number of possible experiments that would have some bearing on our objective is, of course, legion, and we could argue for a long time about which was the right one to do. However, it does not take much thought to expose the inadequacies of the original statement.

For instance, we already know how to grow more food—we can keep twice as many cows or plant wheat on three times as much land. So we have a question: 'More food than what, when or whom?'. That tells us immediately that the statement was useless, simply because we cannot tell from it what constitutes 'more' or how we should know when we had succeeded in producing it. But more critical is that no-one ever means 'more', in the sense of producing it, without also meaning more per unit of something—a hectare, a person, an hour or some other resource.

So suppose we agree that we mean more food (we might even specify tomatoes or dietary energy or milk protein) per hectare of land (per year). Surely this is clear enough? Yes, it is, but it is probably not what we mean. It implies that we wish to know how to produce more food per hectare without regard to the cost of doing so, for instance. Now, we have to be very careful, for the next steps are crucial. Of course we do not mean 'regardless of cost', but what is a tolerable cost? Or do we mean at the lowest possible cost, because, if so, it may not even be possible. It usually turns out that we wish to impose a whole series of conditions: we do not wish to heat the soil, irrigate with nutrient solution each day, enclose the hectare under glass, employ enough people to remove every weed and pest by hand, harvest every single potato produced however hard we have to search, and so on. All of these conditions really have to find expression in a statement of objectives that is fit to base experimentation on and which has any chance of successful achievement.

Very often, a scientist finishes up rather disappointed at the enormous experimental effort that is still required to answer what finally becomes a very limited question, with nothing like the obvious importance that he originally attached to it.

All this thinking about objectives (what is it we really wish to know and propose to spend someone's money finding out) comes before any physical experimentation has been carried out. Or, rather, it should do so.

Having arrived at a satisfactory statement of objectives, many features of the experiment are already and obviously determined, although sometimes broader research and narrower experimental objectives are confused. Thus, if I wish to know something about milk production from cows (a research objective), I might nevertheless use goats or a mathematical model, assuming that the results on one of these will be sufficiently similar to those from cows: but I have really accepted an experimental objective concerned with milk from goats or models, because of my assumptions.

There are other, quite different, decisions to be made, such as the precision with which we want answers. Suppose that we do not know yet how much food (e.g. potatoes) we could produce per unit of two important resources (e.g. land and fertiliser). We may then design an experiment that involves applying different quantities of fertiliser to areas of land and measuring the weight of potatoes produced. We will hope to produce data which can be plotted as a curve on a graph (see Fig. 6.2). This assumes that it is legitimate to join up the points, i.e. that intermediate yields will result from intermediate applications of fertiliser. Now, do we want to know the yield to the nearest kilogramme, gramme or bucketful, and do we want to count *all* potatoes or just the big ones or the undamaged ones and do we want to know their water content or their protein content or whether they shrink on storage or cooking?

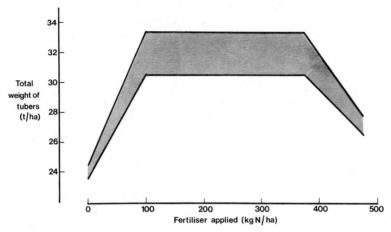

FIG. 6.2.   Response of potatoes to fertiliser nitrogen (after Cooke, 1977).

If we want to be very precise, say to the nearest gramme, then we must also consider how accurately we applied the fertiliser.

These are all commonplace problems of the working scientist engaged in experimentation and they have to do with the kind of information produced by the experiment.

In addition, and quite fundamental to the idea and purpose of scientific experimentation, we need to know whether the results are true, or how we should interpret them. After all, you could experimentally throw a brick at a tree and immediately be struck by lightning: but what is the result of the experiment?

Several important principles have been evolved to guide the experimenter in interpreting the results of his experiments.

First, the experiment is usually designed to test an hypothesis, so the experimenter is saying *in advance* what the experiment is about and what will count as a result and this determines what is measured. Measurement is a characteristic feature, partly because it is helpful to quantify effects but chiefly because it increases *objectivity*. This is extremely important: we are not very interested in whether I *thought* one thing was larger than another, we need to *know* that it was so by some objective test that does not depend on my judgement, eyesight or opinion. Next we need to know whether it was larger because of the treatment being tested. Suppose we grow more potatoes where we have applied more fertiliser: how do we establish a connection between the two?

One common device is to set up a 'control', in which everything is the same except for the fertiliser treatment: if less fertiliser results in less yield, it looks as though there is a positive relationship. But it is very difficult, in fact, to make everything else the same. Perhaps the soil in the control plot has more nitrogen in it, purely by chance. The answer is usually to replicate treatments, so that we do not depend upon one plot of soil, and, by statistical techniques, to estimate the effect of chance, to try and gauge the probability that any result could have occurred by chance.

Thus the methods of experimentation, including design and interpretation, are essential parts of the process by which confidence can be gained that the data collected can bear the meaning that we attribute to them.

However, where natural variation is high—and this is certainly so within agriculture—this confidence may not be very great and it is necessary to repeat experiments in different years, on different soil types, at different altitudes and so on. This is partly, of course, because we cannot control weather and seasons, and, in any case, we may wish to know the answer for a range of weather or even climatic conditions.

'The answer' sounds a perfectly reasonable way of describing the results of an experiment that posed a question, but it is necessary to recognise different sorts of answers.

If we measure potato yields in response to amounts of fertiliser applied, we envisage the answer as a response curve showing the quantitative relationship between yield and fertiliser. In spite of experimental design and statistical treatment, however, this result *might* still be due to chance or might only occur on the sites on which the experiment was carried out. We are, of course, rarely interested in such a result: we usually hope to generalise from these particular results to the whole of England, or to all clay soils or to areas with a given rainfall. After all, if the result of an experiment on a cow only applies to that cow, we have not made a very valuable advance.

It is when we wish to generalise widely that the value of trying to falsify hypotheses becomes evident. The argument here is that we can never be sure that an hypothesis is generally true, because the next test may falsify it. Do we then have to go on and on, and for how long? At least we would have more confidence if we designed the experiment to try and prove the hypothesis wrong: if it still survived intact it would suggest that the generalisation must apply to all less stringent tests.

So a most important experimental technique is to design a situation in which a proposition can be shown to be false, because this only has to be done once to be completely certain. Such a falsification does not, of course, demonstrate that there are no circumstances in which the proposition would be true, only that the generalisation cannot be true.

One of the problems with agricultural research is that, ultimately, it is *particular* solutions to problems that are required: but they have to relate to a particular farmer's field and not just to the particular experimenter's field. This is different, however, from expecting useful generalisations that will apply to *all* fields. Only very simple propositions, such as that more fertiliser (up to some level) will produce more potatoes, are likely to be generally true. The question then is how to find out exactly how much fertiliser a particular farmer should use on a particular field in a particular year. This situation reinforces the view that agriculture is not a science but involves the application of scientific knowledge and principles. There are many aspects of life and human activity that are in this position: indeed, it might be argued that most important human activities involve more than science and that the application of science to such situations represents a major current problem. The fact is that science cannot be applied to non-scientific subjects but the results of scientific activity can certainly be used.

This suggests that other, non-scientific, information and knowledge is also necessary and in agriculture this is so. Perhaps this is true of all situations where decisions have to be taken even though the information available is less than that desired or even that which is adequate.

Since all scientific information is, in a sense, provisional, this is not quite so different and information derived in non-scientific ways can be judged and evaluated in some systematic fashion. Furthermore, we can never expect to obtain all our information by experimentation. Many people object to experiments on living animals and it is not difficult to think of experiments on human beings that we would all regard as intolerable. It is not even possible to experiment in some subjects. Consider the progress made in astronomy, virtually without any controlled experimentation. Consider our confidence in sunrise tomorrow, visible or not, with no basis in controlled experimentation whatever.

A great deal can obviously be learned by observation, generally repeated, of associations between events. This is often the way in which scientific hypotheses are formed and the experimental tests subsequently imposed still depend on observation of their results.

What experiments do is to control the framework within which observations are made and they do this to varying degrees. In general, the practical difficulty is that the more controlled the situation, the less relevant it is, and the more relevant, the less the degree of control.

We then have a choice of trying to understand situations that are 'relevant'—that is, at least similar to the real life situations—but are extremely complex, or of splitting up such situations to deal with bits at a time. If we adopt this analytical approach—and it is, to some extent, inevitable—we must be sure that we can some day put it all back together again. The problems involved in this for agricultural systems were discussed in Chapter 2.

It may then be asked, what is the difference between a systems approach and a scientific approach? Are they in opposition to one another?

The main difference is really in their immediate and direct relevance to real-world problems. Science, in dealing with controlled experiments is generally deriving results from very artificial situations, about which it is relatively easy to state hypotheses. But real-world situations are very complex and, in addition to using science to determine constituent relationships, a systems approach also concerns itself with very complex hypotheses, often only statable as computer models. This capacity to deal with very complex hypotheses, or hypotheses about very complex systems, characterises a systems approach and, in a sense, represents an extension of the scientific method in order to achieve relevance.

One important general point remains to be made. If the hypothesis to be tested scientifically is to be useful in agricultural practice, it must relate to a recognisable agricultural system or to a sufficiently independent part (or sub-system) of one. Furthermore, this relationship must be explicit in the statement of the hypothesis. Now this is more difficult than would at first appear.

For example, try to formulate an hypothesis about a bicycle without actually mentioning the word 'bicycle'. An hypothesis might concern the proposition that whenever the pedals were depressed the rear wheel revolved (or any other proposition you may like to imagine), but it is meaningless without relating it to a bicycle and, even then, including the word only solves the problem if everyone already knows, or can find out, what a bicycle is. (We will ignore, for simplicity in this argument, that there are many sorts of bicycles, having chosen a proposition that probably applies to all of them.) But suppose we are dealing with a system that is not readily denoted by a word (milk production, for example: from what, where, how managed, with what inputs, for what purpose, etc?), then we are in the position of having to describe it in our stated hypothesis. This is difficult enough for a bicycle—just imagine the problems with agriculture! Clearly, the first thing is to aim at a description of the essentials of the system (this is often called a 'model'): the last thing we want is a description that is cluttered up with trivial detail. The next step is to organise these descriptions into a systematic classification.

7

Classification of Agricultural Systems

The value of recognising an individual item as belonging to a class of sufficiently similar items has already been mentioned (Chapter 3). Any information we have about a cow, for example, is vastly more valuable if it applies to all cows, or at least to all cows of the same breed, age or size, or at the same stage of lactation. Information which only applied to one cow would be of very limited value.

Thus research conducted on a few cows has to ensure that they are in some way representative of their class and any record of knowledge so gained has to include details of the class if it is to be used with any confidence.

So scientific research and the communication and use of its results, just as with the fruits of practical experience, really depend upon classification.

However, things can be classified in a great many different ways and there is no one correct way. Classifications must be chosen (like almost all choices) in relation to the purpose for which they are to be used. Classification always involves description, which also has to be related to a purpose and usually involves naming. Names are handy, short descriptions and make it possible to handle complicated classifications. So it is possible to state quite concisely that 'cows and sheep are both ruminants' because (and only if) it is well known what the names mean (ruminants have a special large section of the stomach, the rumen, in which fibrous feeds are broken down by micro-organisms to form digestible components: pigs and poultry are called simple-stomached animals, have no rumen and can use very little fibrous feed—e.g. grass).

If there is no previous knowledge to be 'called up' by the use of names, much more detailed descriptions have to be given. These include the minimum necessary to recognise someone or something: this only needs to include diagnostic features (e.g. the man with red hair, if there is only one such in the room). At the other extreme, if something is to be copied, such as a successful farming system, then all the detail necessary to run it successfully has to be given. It is therefore necessary that this is known, in terms of which detail is essential and which is not—exactly the same problem as in modelling.

Since some propositions will apply to all cows, some to all black cows and some to all one-year-old cows reared in a particular way and fed on silage, classification for any purpose is best done hierarchically, i.e. where each class is part of another and is itself subdivided further (see Fig. 7.1).

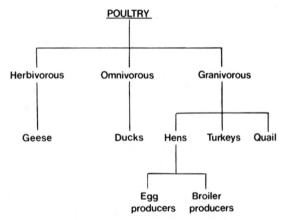

FIG. 7.1.   Hierarchical classification: illustrated for poultry.

Knowledge can then be related to the appropriate level of the classification. If we think in terms of production systems (such as milk from spring-calving cows fed primarily on pasture) within a scheme of worldwide agriculture, then the total classification is bound to be enormous: the general form that it might take is illustrated in Figs. 7.2 and 7.3.

The problems are obvious. Few people wish to spend their time classifying but since it has to be done for clear purposes it cannot just be left to those whose only concern is classification. It is an important activity and it has to be related to an individual's needs, yet we cannot afford the time, nor would it be either sensible or effective for everyone to produce his own scheme.

Of course, it did not happen overnight in those subjects where agreed classification schemes do exist. The important thing is to recognise the need

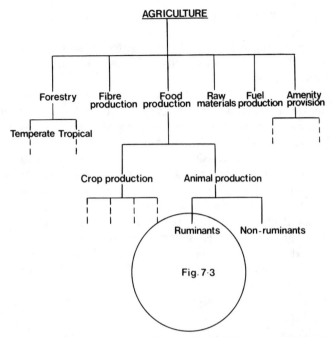

Fig. 7.2.   The classification of world agriculture. (This is, in fact, a rather poor scheme but it serves the purpose of relating to Fig. 7.3 and it is also easy to see alternative ways of constructing the diagram.)

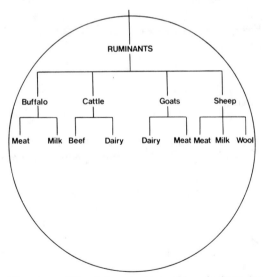

Fig. 7.3.   The place of the dairy cow in world agriculture (see Fig. 7.2).

without exaggerating its priority: it will be better perhaps if a classification of agricultural systems grows steadily rather than rapidly, as the precise need becomes clearer.

The range of possible classification schemes really reflects the variety of ways there are of looking at agricultural systems (see Chapter 3). There are, however, some common features that are encountered in any classification attempt. One is the amount of duplication.

Suppose that we are classifying systems of milk production: it is extremely likely that the criteria we shall wish to use will include the kind of land on which they occur, the part of the world, the climate, the kind of animal and the level of its yield, the stocking rate for grazing systems, the level of capital investment and the intensity of labour use. But these attributes occur in almost all possible combinations, so, whichever is taken first, it is likely that all the other sub-divisions will have to be repeated.

For example, if our first sub-division of world milk production systems is into those in the tropics and those in temperate regions, each of these will have to be sub-divided for, say, level of yield and capital investment (as well as many others).

Thus (Fig. 7.4):

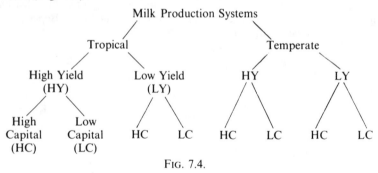

FIG. 7.4.

If, on the other hand, we had started with level of yield, we would have had to divide these into tropical and temperate at some point, thus (Fig. 7.5):

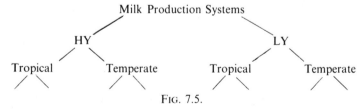

FIG. 7.5.

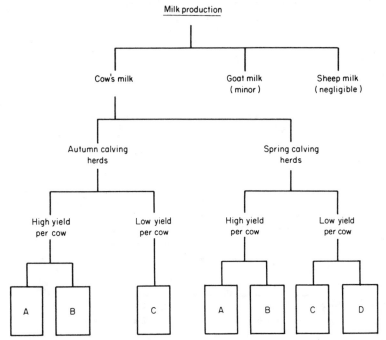

FIG. 7.6. A classification of milk production systems in the UK (A = Intensive concentrate feeding; B = Intensive forage with high yield per cow; C = Intensive forage with high yield per hectare; D = Extensive forage).

Sometimes, criteria only apply to some sub-divisions and do not have to be repeated throughout. This means that it is possible to exercise skill and judgement in placing the most important criteria towards the top of the hierarchy.

An actual classification of milk production systems for the UK is shown in Fig. 7.6. A totally different kind of example, for tomato growing, is given in Fig. 7.7 and a broad classification of world farming systems is shown in Fig. 7.8. The major components of the latter are used as the framework for the discussion of agricultural systems in the next six chapters. These are aimed at a very high level in the classification hierarchy, because it would not be practicable to discuss all of the very large number of systems involved. Within each of the classes considered, therefore, may be included a great variety of systems, all having certain important features in common. The classes differ in many ways: some represent totally different ways of life while others are characterised by the organism used. Some may be found in

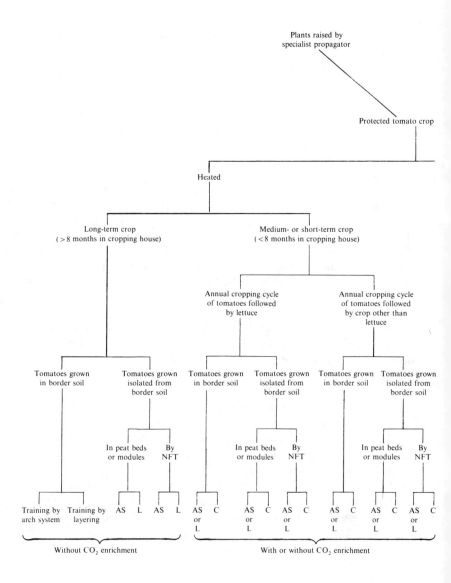

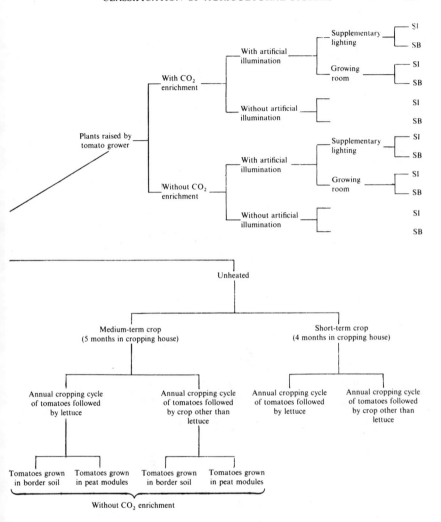

FIG. 7.7. Production methods for tomato crops grown under glass and film plastics in the UK. Methods of pest control, nutrition and watering are not included; the options are similar for most methods of production. NFT, nutrient film technique; AS, arch system of training; L, training by layering; C, conventional training with plants stopped at overhead wires; SI, seeds sown individually in peat blocks or pots; SB, seeds sown in boxes. (Constructed by Dr G. P. Harris, Reading University.)

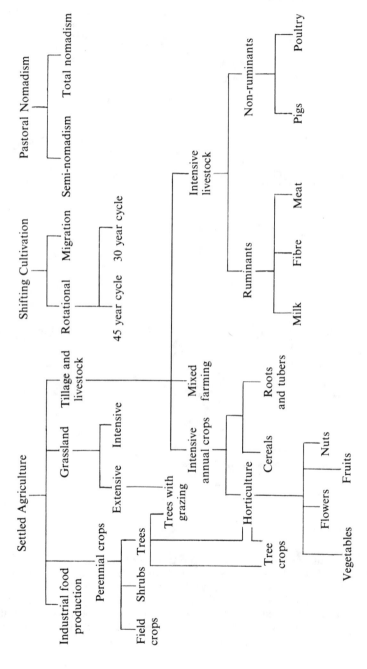

Fig. 7.8. Example of a classification of world farming systems.

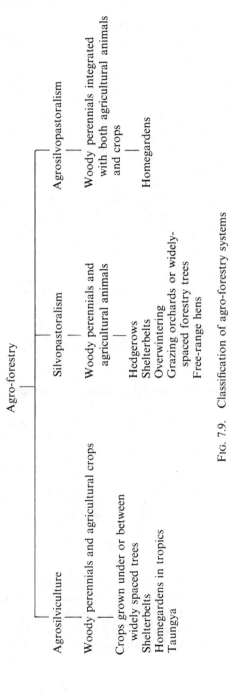

Fig. 7.9.   Classification of agro-forestry systems

the same regions of the world as others, some may only be found in particular regions or at certain stages in the development of a country.

Sometimes it may be possible to introduce systems from one part of the world to another, where their absence may be due simply to the fact that they never developed, or at least to learn from one system features that may be relevant elsewhere.

This was the thinking behind the inventory of agroforestry systems and practices carried out by the International Council for Research in Agro-forestry. Agro-forestry is defined as: systems and practices where woody perennials are deliberately grown on the same land-management unit as agricultural crops and/or animals, in some form of spatial arrangement or temporal sequence, such that the agricultural and forestry components interact beneficially both ecologically and economically.

Such systems (see Chapter 10 for further details) have been classified only in recent years (see Fig. 7.9) and occur in many different forms in different parts of the world.

Agricultural systems occur in relation to some need and are usually constrained by climate and topography.

THE LOCATION OF AGRICULTURAL SYSTEMS

The main factors determining the location of agricultural systems are listed in Table 7.1. It is easier to state them than to decide how they should be measured but, unless this can be done satisfactorily, we are left simply with the idea that these factors may be important. Duckham and Masefield (1970) have suggested that the difference between potential evapo-transpiration and precipitation exerts a powerful influence on the location of farming systems (see Fig. 7.10) and, for temperate areas, have proposed the following formula, to express the way in which the proportion of tilled land is determined by climate (A_k), population density (D_s), economic development (B), access (L_m) and 'local difficulties' (L_0):

$$gA_k - i(L_0 + L_m) + jD_s - kB$$

where g, i, j and k are constants.

The reasons for describing climate mainly in terms of precipitation (P) and potential evapo-transpiration (PET), as in Fig. 7.10, are that plant growth depends upon water supply and solar energy receipt (although other factors may often be limiting, of course). The difference between precipitation and actual evaporation represents a measure of the water

TABLE 7.1
FACTORS DETERMINING THE LOCATION OF AGRICULTURAL
SYSTEMS
(After Duckham and Masefield (1970))

Climate	Precipitation
	Evapo-transpiration
Land	Topography
	Bio-geochemical properties
	Soil stability
Moisture control facilities	Irrigation
	Drainage
Unwanted species	Pests
	Diseases
	Predators
	Weeds
Operational facilities	Tools, machinery, power
Social and economic infrastructures	
Availability of inputs	
Availability of markets	
Feasible crops and animals	

available for plants, although water can also be lost as 'run-off' when rainfall is heavy or when snow thaws rapidly. Evapo-transpiration is also greatly influenced by solar radiation and temperature and is thus a measure of energy receipt. Between the two factors lies much of the climatic influence on crop growth and thus on the location of production systems.

It is unlikely that very precise predictions about location will ever be possible and the factors influencing the location of any particular system will generally be very complex.

The more highly controlled the system, the less sensitive it will be to climate and even cropping systems in the open can be irrigated to mitigate or even eliminate any shortage of rainfall.

The most important aspect of this topic is to be aware of the possibility that any of the factors listed may be a major influence on the location of any of the agricultural systems discussed.

Clearly, crops and animals cannot be produced in unsuitable environments but the latter can be modified to some extent. Soils can be modified—by drainage, irrigation, fertilisers and cultivation—and even the topography can be substantially influenced, in a local sense, by terracing, for example. But, of course, all these modifications cost money or labour and some changes cannot be economically justified.

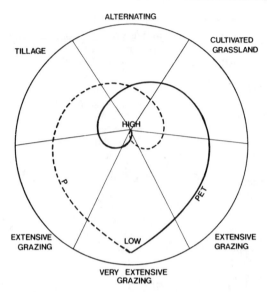

FIG. 7.10. The relations between precipitation (P), potential evapo-transpiration (PET) and the location of temperate farming systems (after Duckham *et al.* 1976, Chapt. 5). Each segment represents the location of each category of farming system and includes the levels of P and PET at which they occur (the centre representing high values and the periphery low values).

Generally speaking, changing light and temperature is impracticable, although greenhouses can be justified economically for very high value crops.

Duckham and Masefield (1970) regarded the main 'influents' on location as: (a) access to mass markets; (b) the size of the local population; (c) the level of social and economic development; (d) various 'operational factors' (such as suitability of the terrain for mechanisation) and (e) climate. These 'influents' (*i*) were thought of as affecting resultants (*r*) which in turn determined location: thus climate (*i*) affects soil and vegetation (*r*) and the size of the local population (*i*) affects the local demand for farm products (*r*).

The classification of farming systems that resulted from all this started with four main classes: (i) plantation perennial, (ii) tillage, (iii) alternating between tillage and grassland and (iv) grassland or grazing. The last class can occur in a very extensive form, or less so, or as intensive 'cultivated' grassland (i.e. sown leys).

The relationship between the location of most of these and two major factors—precipitation and potential evapo-transpiration—is what is illustrated in Fig. 7.10. With high values in the centre and low values at the periphery, the lines for precipitation (P) and potential evapo-transpiration (PET) show that extensive grazing occurs where one of these is high but the other low. Tillage, on the other hand, is limited to areas with high PET and moderate P, whilst for cultivated grassland the position is reversed and alternating systems occur where both P and PET are moderately high.

Of course, many other factors may be of great importance, too. In the tropics, for example, day-length is nearly constant and temperatures tend to be high, although the latter also depends upon altitude. However, the absence of a cold winter and plentiful solar radiation may have adverse effects in terms of pests, weeds and diseases, as well as beneficial effects on crop growth.

Sometimes the climate will allow two crops a year and, where no frosts occur, areas may have a monopoly of certain crops. Some hazards (such as floods, hurricanes, severe droughts and locusts) may also be associated with these areas, however, and the occurrence of high winds and hurricanes may rule out tree crops as being too risky.

Tropical soils are generally regarded as poor and with low organic matter due to rapid degradation of such material. This is by no means always the case, however.

Soil fertility is much influenced by leaching and this is related to the frequency of excessive rainfall. When rainfall is so heavy that it cannot be absorbed by the soil, 'run-off' may occur to such a degree that serious soil erosion occurs, often by the formation of gullies or even landslips. This is common in the wet tropics, although not confined to them, and erosion is a serious problem in countries like New Zealand, for example.

An important aspect of climatic factors is that their seasonality may be even more important than their annual totals. Thus, average annual precipitation does not tell you how much falls as rain or snow, on any one day (or in any one hour), how long a period there may be with no rain at all and how large is the variation between years. Seasonal variation in day-length means that short season crops may receive more solar radiation in temperate zones than in the tropics but that perennials may do better in the wet tropics. In the seasonally dry tropics, perennials have sufficient solar radiation throughout the year but there are seasons when the rainfall is inadequate for growth.

Similar considerations apply to livestock production systems. They depend upon the crops or vegetation that can be grown and are further influenced by direct effects of climate (especially temperature and

TABLE 7.2
WORLD GRAZING AREAS LIMITED BY TEMPERATURE OR MOISTURE
(After USSAC, 1967)

Limitation	Grazing area (Mha)
Temperature	928
Moisture	1 456
Temperature and moisture	912
Neither temperature nor moisture	324
Total	3 620

insolation), by the nature and quantity of the feed available and by the incidence of pests, parasites and disease.

Whatever generalisations are made about the location of agricultural systems or about their classification on any ecological basis, it is clear that there will also be found enormous local variation within ecological areas. Just as you *could* have a glasshouse at the South Pole (and they do have them, heated by geysers, in Iceland), you *could* have intensive poultry houses in the dry savannas: whether you did or not would clearly be determined by a whole range of factors.

On a world basis, the main limitations on agricultural production are temperature and rainfall.

By way of illustration, Table 7.2 shows the estimated grazing areas that are limited by either or both of these two factors.

8

Subsistence Farming and Shifting Cultivation

Subsistence farming simply implies producing enough food and fibre for the needs of the farmer and his family. At one time, not so long ago, this was quite common, especially in the tropics. People worked for relatively few hours each week, produced only what they required, collected fuel from the surrounding areas and played little part in the cash economy.

Today, it must be relatively rare to produce nothing for sale at all but this may nevertheless fall far short of a commercial attitude to farming. Of course, commercial farming requires markets and an infrastructure of roads and other facilities. Peasant farmers, their wives and children, often have to carry very heavy loads and drive animals for many miles in order to make use of a market. All this may be time-consuming and laborious but it may also be part of their way of life, with many social and cultural connotations.

Shifting cultivation is also a primitive form of agriculture, although there is no reason why it should only support subsistence farming. As the name implies, it involves periodic shifts to a new piece of land, the fertility of the original patch having been exhausted.

Any form of crop production must remove elements from the soil and these have to be replaced. Growing legumes may replenish the nitrogen but potash and phosphate have to be returned to the soil in one way or another. If they are not, yields will fall (they may do so because of a build up of pests and disease as well, or, more likely (Webster and Wilson, 1966) a deterioration in soil structure) and the only solution is to move to a fresh piece of land.

This commonly involves clearing forest or bush and the vegetation may be burned. The resulting ash serves as an initial fertiliser and, in some systems, such as the 'chitamere' system of Zambia and the 'hariq' system in the Sudan, branches or grass are collected from a large area to be burned on the newly cleared land.

Of course, tree stumps and boulders may limit the amount and type of cultivation that can be practised (as may termite mounds), although with plenty of labour it is perfectly possible to cultivate around obstacles.

Shifting cultivation may move on rapidly or slowly and may return to the same areas at greater or lesser intervals. Regular rotation may be practised (Duckham and Masefield (1970) refer to the rotation of cleared land, with between one-quarter and one-third being cropped with cotton, winter wheat, oats, hay and peaches). As population increases, the length of rest period diminishes and soil exhaustion and soil erosion may follow, as in parts of East Africa. Where population density is low, as on the Pacific coast and in Colombia, shifting cultivation can avoid these dangers.

According to Nye and Greenland (1960), shifting cultivation occupied over 200 million people on more than 30 % of the world's exploitable soils. It is therefore a major form of land use, practised by people in New Guinea, Asia, Africa and Latin America, in a variety of forms.

These were discussed in some detail by Ruthenberg (1971) for the tropics: he distinguished the following systems:

Vegetation systems (bush, forest, grassland)
Migration systems   (random, linear or cyclic shifts)
Rotation systems   (45 years or 30 years)
Clearance systems   (burning, hoeing, cutting, etc.)
Cropping systems   (rice, root crop, maize, millet)
Tool systems   (axe, matchet, digging-stick, hoe, plough)

A different classification was used by Okigbo (1977). This distinguished first between the 'traditional and transitional' farming systems of tropical Africa and the 'modern systems and their local adaptations'.

Amongst the first category, Okigbo listed four phases of shifting cultivation and 'terrace farming and floodland agriculture'.

The four phases of shifting cultivation were characterised by decreasing length of fallow periods, increasing cultivation of grain legumes and a move towards rudimentary sedentary systems and compound farms or homestead gardens, often associated with outlying farms with marked fallow periods. The homestead gardens are intensively cropped, with or without reduced fallow periods, and the crops tend to be grown in a mixed culture of annuals, biennials and perennial trees and shrubs.

In fact, one of the characteristics of shifting cultivation systems is the use of mixed cropping, i.e. two or more crops grown intermingled, not in adjacent rows (intercropping). The advantages of this (see Chapter 10) are a reduction in susceptibility to pests and diseases and a better use of the environment. Of course, the latter is only the case where species are grown that have different light requirements or explore different depths of soil. Mixed cropping also tends to provide a complete vegetation canopy, although at different heights, and thus breaks up heavy rainfall, protecting the soil.

In hot, humid climates, soil fertility and, consequently, crop yields, decline rapidly, in a matter of two or three years. This is inevitable in the total absence of fertiliser but shifting cultivation need not depend for soil fertility solely on the regenerating effect of a forest or bush fallow. Hut refuse, green manuring, compost, ashes from tree and shrub burning and some animal manures may be used: purchased mineral fertilisers are very little used.

The availability of animal manures depends on what animals are kept. In areas infested by the tsetse fly, cattle rearing is limited to those, such as parts of West Africa and the Congo, where trypano-tolerant breeds occur. In the rainforests, only a few goats, sheep or chickens are kept: in south-east Asia and West Africa pigs are reared.

In Latin America, however, substantial herds of cattle are often combined with shifting cultivation. Wherever animals are confined, even for periods only, manure can be collected for use on the cultivated areas.

CAPITAL AND LABOUR IN SHIFTING CULTIVATION SYSTEMS

Capital is usually low, except where animal numbers are high.

The labour input is primarily in the form of manual work for clearance, the preparation of land, planting, weeding, bird scaring, harvesting and processing. Ruthenberg (1971) has illustrated the great variation in labour input for clearance, depending on the system employed (or needed).

Slashing and felling may only require 42 man-hours per hectare on grassland, for example, whereas the figure may be as high as 492 for rainforest. Burning and clearing, however, may take 780 man-hours per hectare on that same grassland (including hoe-cultivation) but as little as 32 for secondary forest (in Borneo). The totals for all these operations varied from 822 for grassland in the Congo to 486–1450 for Congo rainforest.

Work in the fields occupies probably no more than half the working time, however, and tending livestock may be a significant part of the remainder.

As already mentioned, trips to the market, collecting fuel and fetching water are also very time-consuming.

There is usually a distinct division of labour between men, women and children, with men doing the heavy clearance work.

Improvements can be introduced in order to reduce the element of drudgery; better tools and different crops can lead to higher yields. But this may only increase the period required under fallow and even reduce the period of cropping.

The main developments from shifting cultivation are either to produce cash crops or to evolve into quite different forms of land use.

THE IMPROVEMENT OF SHIFTING SYSTEMS

The fallow period is usually one of restoration of fertility by the regeneration of trees and bushes. It is possible to speed this up by planting rapid growing species or legumes, or to combine shifting food cultivation with foresty (as in India, Burma, Indonesia and East Africa).

Shifting systems have also been combined with plantations in Sumatra, but progression to settled, intensive crop production depends upon an input of mineral fertiliser.

A progression to grassland is possible, however, and, in Brazil for example, shifting cultivation has been used as a method of extending the area of cleared grazing land.

Evolution *can* occur to perennial crops and often does so into tree crops, such as rubber, cocoa and coconuts for sale and banana groves for food. In the warm, damp climates of tropical Asia, paddy-rice systems may develop.

Clearly, shifting cultivation is an obvious way for early agriculturalists to use land in a climate where the regeneration of natural vegetation is rapid. Nevertheless, it involves a lot of initial clearance work that has to be repeated at intervals of 3–4 years and, because of this, there are limits to the extent to which clearance is worth while. The removal of roots, tree stumps and boulders may not be worth while, neither drainage nor irrigation can be justified, paths are no more than footpaths, thus limiting the forms of transport, and, periodically, the hut site has to be moved and a new hut built.

All this naturally engenders attitudes of just doing enough, rather than encouraging any attempt at steady improvement. The latter really depends on some degree of permanency.

There is thus a similarity between the way of life of the shifting cultivators and that of the pastoral nomads discussed in the next chapter. However, whereas the nomads have to cover considerable distances to feed their animals, the cultivators shift their shorter distances largely because of a decline in soil fertility that is not, in fact, inevitable.

After all, apart from what is sold, nothing need be lost to the soil. If all household waste and animal manures were carefully husbanded and returned to the soil, intensive crop production could be practised indefinitely.

Subsistence farming cannot be regarded as an end in itself, unless of necessity. Every household needs to purchase something (implements, medical care, education) and must therefore produce more than is consumed. Subsistence farmers are very vulnerable, with little to insure them against climatic or other disasters, but very often societies evolve an elaborate network of family relationships that provide help in difficult times.

Small farmers, in general, tend to be vulnerable, having less credit, little bargaining power and few reserves or resources. They are therefore often exploited by middlemen who buy their products and by suppliers of goods and services.

Relatively little research has usually been directed to the needs of small farmers and the results of research designed for large farmers may be quite inappropriate. The Intermediate Technology Development Group (ITDG) exists to develop and supply appropriate technology to small farmers and research centres are beginning to re-orient their programmes.

Even so, the difficulties of small farmers are considerable. They cannot all become big farmers and they compete with each other in the market place.

One important solution is the formation of farmer co-operatives. Where these are most successful, they cover the whole range of activities, buying from the farmers, selling to the consumers and supplying inputs back to the farmers.

But small farmers may not have the knowledge, experience or skills to run co-operatives—including simple book-keeping, for example. Sometimes, the most effective form of help would be in this direction, and nothing to do with technical matters at all.

Whatever developments are proposed to improve the lot of the small farmer, it is important not to destroy social structures that have evolved over many years and on which people depend to get them over difficult periods. If you only have one cow, the death of a cow can be catastrophic. If your area of grassland is burnt up by drought, access to that of neighbours may sustain your livestock until the rains arrive.

Social infrastructures are often of very great importance in pastoral nomadism, dealt with in the next chapter.

9

Pastoral Nomadism

Nomads are people who travel more or less continuously, with no settled homes, although often following well-established, traditional routes.

In the arid and semi-arid tropics, the yield of grassland areas is extremely low and highly seasonal: it is only possible to live off a very large area. If such large areas are owned they can be ranched but, apart from this, the only alternative is nomadism.

Nomadism may be total† or partial, varying with the extent to which farmers live in permanent settlements. Total nomads have no permanent homes and do not practise any cultivation, moving with their herds of livestock. However, permanent residence and some cultivation can be combined with long periods of travel with the herds (semi-nomadism) and this gives rise to 'transhumance', in which farmers with a permanent home go (or send a herdsman) with their herds to distant grazing areas for substantial periods of time.

Examples of nomadic tribes are the Masai of Kenya and Tanzania and the Hima of Uganda, who live on meat, milk and blood from their cattle, the Fulani of West Africa, the Bedouin of the Eastern Mediterranean and the Mongolian nomads.

It is difficult to assess the importance of pastoral nomadism. Although it is difficult to obtain data on a world basis, especially, for example, in relation to China, it has been estimated by Grigg (1974) that the total number of pastoral nomads probably does not exceed 15 million (see Table 9.1). However, it has also been estimated that they occupy some 10 million square miles of the earth's surface—nearly twice the world's cultivated area. Even on a simple numerical basis, the nomads are of considerable

† See Ruthenberg (1971) for a more detailed treatment of this subject.

TABLE 9.1
NUMBERS OF PASTORAL NOMADS
(After Grigg, 1974)

Region[a]	Numbers[a]
Sahara	c 1 000 000
Sahel and the Sudan	2 500 000
Syria, Jordan and Iraq	650 000
Iran and Afghanistan	5 000 000

[a] Both are very approximate and the available information does not account for anything like the estimated total.

importance in particular countries (see Table 9.2) and they may own an even higher proportion of the livestock.

The latter are chiefly camels, cattle, sheep and goats and the products include milk, blood, meat (chiefly consumed on ceremonial occasions), hides, wool and skins.

Nomadic herders rely solely on natural pasture, although this includes trees, such as *Acacia* spp., that provide fruit, leaves and even branches for feed, and they do not even store reserve feeds or forage. They are continually moving in search of pasture, water and, sometimes, salt.

Pastoral nomads are sometimes classified as 'vertical' or 'horizontal' according to whether they migrate up and down mountains, with substantial variation in altitude, or laterally around areas at roughly the same height. In both cases, the routes are chosen with regard to the seasonal availability of pasture and water.

The productivity of these pastures is low but depends, of course, on the

TABLE 9.2
NOMADS AS A PROPORTION OF THE POPULATION
(After Grigg, 1974)

Country	%
Somalia	c 75
Sudan	30–40
Iran and Afghanistan	15–20
Saudi Arabia (in the 1950s)	c 50

rainfall to a very great extent. Population densities are therefore very low: typical estimates range from 0·5 per square kilometre in the Sahara and the Sahel to 17 per square kilometre in East Africa.

Nomads usually migrate in groups of five or six families, each requiring anything between 25 to 60 goats and sheep or 10 to 25 camels. Sheep and goats tend to predominate but camels, dromedaries and horses are the prestige animals and confer great mobility.

Productivity in these systems tends to be very low, although not necessarily lower than that of other systems in the most arid areas without irrigation. However, even on the same soil type and under the same climatic conditions, ranching would be expected to produce a sale of about 20% of the animals annually, whereas those involved in semi-nomadism would only slaughter and sell about 6–10% of their stock and these would be of the poorer quality. The disadvantages of semi-nomadism have to do with the combination of individual animal ownership and communal grazing and the control of disease. Common grazing generally has the effect of over-grazing, since every individual farmer wishes to maximise his number of animals.

If nomadism spreads it can increase the areas of severe over-grazing and it is in this sense that the tsetse fly has sometimes been regarded as the most important agent for conservation in Africa.

The same arguments can be applied to other measures of 'improvement'. For example, the provision of more watering places can result in an increase in over-grazing, since it makes it possible to keep a higher number of animals than the original water supply would have allowed. Curiously enough, semi-nomadism may be more damaging, because it occupies areas that could sustain commercial farming and dissipates resources needed for such developments.

Total nomadism usually occurs on land that could sustain no other form of production and may well be the best way of exploiting the arid steppes and desert (see Fig. 9.1). The problems are that there is little scope for technical improvement and the nomads find it difficult to compete in a world that is changing all around them.

As their own numbers increase, the grazing areas available to them diminish and their incomes decrease (due to competition with other forms of transport, e.g. motor vehicles, and with urban traders). On top of this, in years of severe drought the twin crises of gross over-grazing and starvation produce intolerable conditions. In many parts of the world, therefore, considerable efforts are being made to persuade the nomads to adopt a settled way of life.

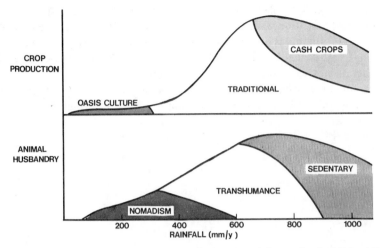

FIG. 9.1.   The relationship between agricultural activity and rainfall in Africa
(after Matlock and Cockrum, 1974).

Of course, the nomads' animals are themselves a mobile food reserve and
there are long-established cultural practices (such as 'stock-friends'—
raiding and redistribution via ritual leaders) for meeting the special needs of
individuals at critical times. None of this can cope with major or prolonged
drought, however, and in some regions (e.g. the Sahel) the situation has
reached desperate levels in recent years. Major studies of the process of
'desertisation' have been carried out and have identified over-grazing, over-
cultivation, excessive wood collection and bush burning as the practices
that lead to degradation and eventual desertisation of dry rangelands.

However, the final conclusion of one recent study (Rapp et al., 1976)
illustrates the difficulties: halting desertisation and restoration of degraded
land requires a strong political will, the power to implement the necessary
measures, co-operation with the local population and employment as an
alternative to emigration.

The fact is that there are no simple solutions to any of these problems and
those who press for settlement of nomads, reduction in the numbers of
livestock, decrease in the growth of human populations and so on, each
has a point, but none of these measures by themselves represents the
answer.

Desertification is certainly a major problem in many parts of the world
and deserts can be created or extended in a number of different ways.

Major projects to plant trees or other vegetation, in order to prevent further desertification, face formidable difficulties. There is a tendency to plant vegetation that is useful—to people for fuel or to animals for forage—but this usually means that it is unlikely to survive: and hungry people and animals do not wait for the vegetation to grow.

It is for this reason that some people have argued for a 'green glue' of poisonous, spiny plants to combat erosion, because such plants are more likely to survive.

10

Mixed Farming Systems

Few farming systems confine themselves to one enterprise, producing one main product. Most are mixtures of several, often quite different, enterprises, so it would seem that a very high proportion of the world's agriculture could be said to be engaged in mixed farming. This is not a very helpful view, however, and it is more useful to confine the term to situations where crop and livestock production are integrated.

Integration can nevertheless take many forms and, indeed, the inherent flexibility of the system is one of its major advantages.

In some cases it is primarily the labour force that is integrated (see Fig. 10.1 for an example) and enterprises are chosen in relation to when their peak labour demands occur. Most commonly, there is also an integrated land-use policy, over a period of time. This is reflected in rotations, to the benefit of the land and its fertility, the control of both crop and animal pests and diseases and the suppression of weeds. Much closer integration also occurs with shared use of land within a year, as in alternate grazing by cattle and sheep, or mixed grazing by more than one animal species simultaneously. Similarly, many of the crops grown may be solely for feeding animals, either directly—as with grass and forage crops for grazing—or after harvesting and, often, storage, as with barley, maize and roots.

The origins of mixed farming could obviously have to do with any of these facts but were probably also influenced by the need to keep and feed draught animals and the advantage of producing high value livestock products for sale.

Grigg (1974) argues that mixed farming, as practised in North America and western Europe, can be traced back to the early village sites of

125

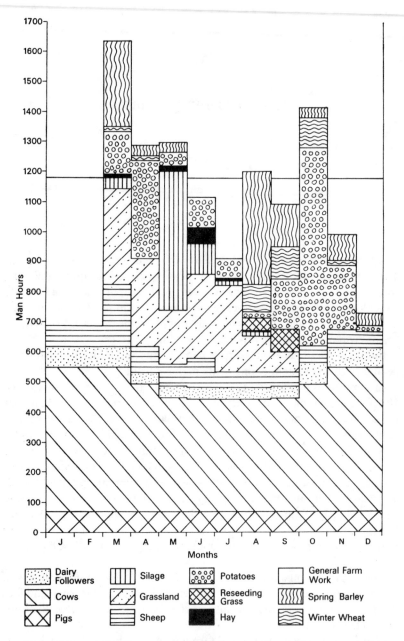

FIG. 10.1.   Integrated labour use in mixed farming (from Norman and Coote, 1971. Reproduced by courtesy of Longman Group Ltd).

south-west Asia. He identifies the significant moments in the evolution of modern mixed farming systems as:

(1) The introduction of the heavier plough and the three-field system in the early Middle Ages.
(2) The reduction of fallow and the growing of roots and grass as fodder.
(3) The intensification of farming and the shift to livestock products since 1850.

These developments have often taken place close to urban areas and have depended upon the availability of manufactured inputs, such as machinery and fertiliser, and on an outlet for the products, very often to processors rather than to consumers directly. The latter point is of some interest. It is well exemplified by butter and cheese, once made on the farm but now only rarely so in developed agricultural systems. This leads to a situation where farming may not produce food so much as raw materials for a food industry. It is worth remembering at this point that most farm products that are destined for food always have undergone quite a lot of preparation and processing, including cooking, but generally by the consuming household.

The distinctive features of modern mixed farming systems are that they are highly commercialised with trends to increasing mechanisation and a reduction in the labour force. The importance of the last point varies with farm size, however, and a high proportion of the farms of north-western Europe and the eastern part of the United States are characteristically based on family units with little hired labour.

The degree of agricultural integration within these farms varies but, typically, a variety of crops is grown, generally including grass, cereals and, in Europe, roots. Typical cereals are oats, barley, wheat and rye, with maize in the Corn Belt of the USA. Common root crops are turnips, mangolds and swedes for fodder, and potatoes and sugar beet, the former for direct consumption, for feeding pigs and cattle and for sale to distilleries (in Germany). This variety reduces risk but also evens out the labour requirement.

The larger part of the income is derived from livestock products and a high proportion of the grain produced is fed to livestock.

Productivity of mixed farming tends to be high per unit of land but not per man: this, of course, is also related to farm size and the possibilities of economies of scale. The advantages of mixed farming are primarily a reduction in risk and better use of total resources because of the integration of enterprise needs that is possible. However, this same complexity also makes greater demands on management skills and the husbandry knowledge and marketing expertise required, and may place undesirably low limits on the size and scale of each enterprise. It is not surprising, therefore, that a farmer should look at his enterprises, decide which is the most profitable and, to a greater or lesser extent, specialise in that one.

Specialisation carries its own advantages and risks but, to be possible at all, some elements of the mixed farming situation have to be 'dispensable'.

Thus, for example, crop rotations designed to control pests or weeds may be unimportant if pesticides and herbicides are available at a price that makes economic sense. Animal manures may similarly be dispensed with if artificial fertilisers can replace them economically.

In these examples, additional inputs are needed that require energy in their manufacture. For this and other reasons, the price of such inputs may rise to a point where it is more economical to revert to some features of mixed farming.

THE VALUE OF MIXED FARMING

The difficulties of assessing the real value of mixed farming are, however, considerable. First, there is no one version that is in any sense representative, so there is no way in which 'mixed farming', as such, can be unequivocally good or bad. Secondly, the value of such features as rotations, and their effects on soil fertility, may take many years to find expression and results and experience may be confounded by variable weather and changes in management. Thirdly, it is a subject on which controlled experimentation is virtually impossible.

A systems approach is thus almost essential in thinking about mixed farming. One only has to consider how one would set about an experimental study of such situations to realise the difficulties: the result is that scientific findings are tried out on farms, with all the problems of interpreting the outcome. Such 'trial and error' tests are extremely inefficient. (Imagine this approach to, say, launching a rocket to the moon!)

The complexity of mixed farming systems can be represented in models and some experimentation carried out in this way. Essentially, the rotational element of mixed farming serves the same sorts of purposes as the movement to new land involved in both shifting cultivation and pastoral nomadism. The ability to do without this element depends largely on the availability and cost of the necessary additional inputs and this is something that is not likely to be constant over either time or space.

The advantages of integration are also affected by availability and costs of inputs since integration may be most useful where control is less than total. The greater the degree of control, the less the need to integrate, because the latter is concerned with balancing and combining different cycles, peaks and troughs, whereas control can eliminate or determine them. Thus, in a completely controlled environment there need not be phases in production, seasonality of growth, reproduction, labour or any other resource need. All these can be adjusted to desirable patterns, but such control may have a high cost.

Whatever the integration planned or achieved, however, most enterprises are based on relatively separate crop or animal production systems. These are dealt with in the next two chapters.

The major exceptions include agro-forestry systems. As defined in Chapter 7, agro-forestry consists of both agricultural and forestry components occupying the same space and time, for both ecological and economic benefit.

A good example of such a system is the alley cropping of Leucaena in Southern Nigeria (described by Ngambeki, 1985). This system uses the leguminous Leucaena hedges to supply controlled (by pruning) shading and green manure mulches to the benefit of companion crops of maize or cowpea. In the case of maize, yields can be increased in this way, without the use of nitrogenous fertiliser.

Another example is to be found in tropical homegardens (as described by Fernandes and Nair, 1986). These contain multipurpose trees and shrubs in intimate association with annual and perennial agricultural crops and, invariably, livestock. They are found within the compounds of individual houses, the whole tree–crop–animal unit being intensively managed by family labour. Such homegardens are often small (<0·5 ha) and are found in all ecological regions of the tropics and sub-tropics, especially in humid lowlands with high population density.

11

Crop Production Systems

Agricultural plants may perform a variety of functions, including the provision of shelter for Man, animals and other plants, the control of soil erosion, to serve amenity purposes (lawns, flowers, trees and shrubs) and to increase soil fertility (e.g. by adding organic matter from senescent leaves and roots or nitrogen from decaying legume nodules).

The major crop production systems, however, are designed to produce food for direct human consumption, feed for animals, fibre for fuel, construction or manufacturing, and a miscellaneous group of products such as tobacco, perfumes and drugs.

In terms of the land area used, seven crops occupy 71 % of the cultivated area of the world (Table 11.1). These are all grain crops and their importance is further indicated by the fact that they contribute substantially to the energy intake of the world's population. The other crops that make major contributions to dietary energy supply are potatoes ($c.$ 5 %) and cassava ($c.$ 2 %).

On average, world energy intake is derived mainly from grains, roots and tubers (totalling 62·7 %), fruits, nuts and vegetables (9·6 %), sugar (7·3 %) and fats and oils (8·9 %). Livestock products (10·8 %) and fish (0·7 %) are relatively unimportant on average but livestock products may represent a high proportion (up to 35 %) in certain economically developed regions: equally, they may fall as low as $c.$ 5 % in less economically developed regions where grains, roots and tubers may exceed 70 %.

Since the environments in which these crops are grown vary enormously and the levels of crop protection, fertiliser input and husbandry also vary greatly, it is not possible to state the yields of these major crops. They vary in time and space and only ranges can be given (Table 11.2). Protein content varies less (Table 11.3).

TABLE 11.1
HARVESTED AREAS OF WORLD CROPS (AS % OF TOTAL
ARABLE LAND)
(After FAO, 1985a)

Crop	%
Cereals	52·0
Wheat	16·8
Rice	10·5
Maize	8·6
Pulses	4·8
Roots and tubers	3·4
Sugar	1·8
Cotton	1·0

Yields of energy vary with crop yields, of course, but there are considerable species differences in potential output per unit area of land (Table 11.4).

Included amongst the foods are the 'beverage' plants (those giving rise to coffee, tea, maté, chocolate, cocoa, alcohol and wines), those producing flavourings, spices and the vegetable oils (cottonseed, maize, oil palm, olive, peanut, safflower, soybean). Non-food products include waxes, pectins, gums, insecticides, drugs and, above all, fibres. The latter include soft fibres (such as flax and hemp), hard fibres (such as sisal) and surface fibres (like cotton, kapok and coir).

A different type of fibre production is in the form of wood, from a whole range of hard- and softwood trees. By comparison with the land area that is cultivated, the area under forest and woodland is very large indeed (see Table 11.5) and its net primary productivity is relatively high.

TABLE 11.2
EXAMPLES OF YIELDS OF THE MAJOR CROPS
(After Cooke, 1983)

	Yield ($t\,ha^{-1}$)			
	Wheat grain	Rice grain	Maize grain	Potato tubers
Established potential yield	14	25	22	103
Average yields recorded by FAO				
All developed countries	2·1	5·3	5·4	15·6
All developing countries	1·7	2·8	2·0	10·8
Highest country average	6·7	6·1	7·9	39·0
	(Netherlands)	(Korea DPR)	(Greece)	(Netherlands)

TABLE 11.3

PROTEIN CONTENT OF THE MAJOR FOOD CROPS

	Crop	Crude protein (% of DM)
Grain	Wheat (flour)	14
	Rice	8
	Maize	10
	Barley	10
	Sorghum	11
Oilseeds	Rape	18·9
Roots	Turnip	12
	Sugar beet (root)	2·6
Tubers	Potatoes	7–9
Pulses	Beans	28–30
	Chickpeas	20
	Soyabeans	43
	Groundnut	26
	Peas	28
Forages		6–25
Fruits	Grapes	1·44–2·45
	Bananas	1·2
	Apples	0·6
	Citrus	0·5–1·7
Sugar cane		3·8
Vegetables	Cabbage	15·0
	Carrot	7·3
	Spinach	36·0
	Tomato	15·0

TABLE 11.4

GROSS ENERGY OUTPUT (MJ/ha/y) FOR THE MAJOR CROPS UNDER 'NORMAL' CONDITIONS (i.e. WHERE THEY ARE NORMALLY GROWN) (Sources: FAO, 1985a; Spedding, 1984)

Crop	MJ/ha/y
Maize (grain: North America)	114 000
Wheat (grain: Netherlands)	124 000
Barley (grain: Netherlands)	89 000
Rice (grain: Japan)	99 000
Potatoes (tuber: Israel)	170 000
Perennial ryegrass (harvestable crop: UK)	222 000
Cabbage (harvestable crop: UK)	105 000
Sugar beet (harvestable crop: Greece)	229 000

TABLE 11.5
THE VEGETATION OF THE WORLD

Vegetation	Area (10⁶ km²)	Growing season (days)	DM yield potential (t/ha/y)
Forest	50·0		
Evergreen			
Tropics with high humidity		365	146
Tropics with bimodal rainfall		260	104
Tropics with monomodal rainfall		180	72
Temperate		180	52
Deciduous			
Tropical		180	37–56
Temperate		180	46
Woodland	7·0		
Dwarf and open scrub	26·0		
Grassland	24·0		
Temperate			16·7–26·6
Sub-tropical			23·5–31·9
Tropical			48·8–85·2
Desert	24·0		Negligible (currently)
Cultivated land	14·0		
Tropics			
Maize		260 (2 × 130)	37
Soyabean		260	21
Temperate			
Winter cereal		319	26
Spring cereal		152	21

FEED FOR ANIMALS

Many of the food crops grown for direct consumption can be, and are, fed to animals. Grains may be fed to poultry and pigs, some cattle-raising systems are based on grain and small amounts of grain are fed to sheep. Animals are also fed on waste and by-products from crop production systems.

Amongst the crops grown specifically to provide animal feed (Table

TABLE 11.6
CROPS AND BY-PRODUCTS GROWN FOR ANIMAL FEED

Crops
  Concentrates   Cereal grains, pulses, oilseeds
  Fresh green feed   Grasses, forage legumes, brassicas
  Bulky conserved feeds Hay, silage, dried crops
  'Roots'   Potatoes, swedes, turnips
Crop Wastes and By-products
  Sugar beet tops
  Bean and pea pods and haulm
  Straw
  Chaff
Industrial Wastes and By-products
  Sugar cane bagasse
  Sugar beet pulp
  Brewer's grains
  Blood, meat and bone meals
  Fish meal
  Bran
  Meals from extracted oil seeds
  Citrus pulp

11.6), grassland occupies by far the largest area of land (see Table 11.7) and sustains the greatest number of animals.

Grassland systems really belong amongst animal production, since their output is usually in the form of animal products. This is not inevitable and dried grass and hay may be sold, for feeding by others, in the same way as is barley, for example, but this does not occur to any great extent.

TABLE 11.7
LAND AREAS USED FOR FEEDING ANIMALS
(FAO, 1977)

	1000's ha	Percent of land area
Total land area	13 075 336	100
Arable land	1 415 467	10·8
(some of which also produces animal feed)		
Permanent crops (e.g. coffee, rubber)	90 672	0·7
(some animals fed on by-products)		
Permanent pasture	3 046 404	23·3
(entirely used for feeding animals)		
Forest and woodland	4 156 355	31·8
'Other' land (towns, wasteland, etc.)	4 366 428	33·4

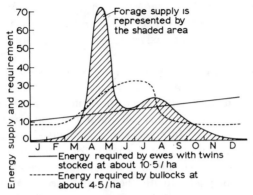

FIG. 11.1.   Distribution of feed supply and requirement for sheep and cattle (after Spedding, 1971). The stocking rates are calculated to match the total forage supply (5800 kg of digestible organic matter per hectare per year).

One of the reasons for the specific production of animal feeds is that the distribution of plant growth is seasonal and does not normally supply material in the same quantities as are required by warm-blooded animals (Fig. 11.1). The latter need feeding at least as much when the temperature is too low for plant growth and, although their needs are not constant, they have no periods when their needs are negligible. Some of the plant growth therefore has to be stored (usually as hay or silage), or supplements (often cereals) have to be added, or special crops have to be grown (such as rape, kale and roots). In many extensive grazing systems, the major fluctuations in plant growth have to be met by animals living on their reserve fat during periods of feed shortage, renewing these reserves when feed is plentiful.

There are some special features of grass production systems but these will be referred to later in this chapter.

Productivity is largely determined by the environment, whether natural or managed artificially.

THE ENVIRONMENT OF CROP PRODUCTION SYSTEMS

The essential physical features of the environment are the soil, as a rooting medium and a source of water and mineral nutrients, and the atmosphere, as a source of oxygen, CO_2 and nitrogen, and as a 'sink' for water vapour and gases, such as O_2 and CO_2, that are released, usually via the stomata.

These physical features are influenced by climate and weather, and crops can be greatly affected by rainfall and wind severity, by temperature and by solar radiation directly. The latter is the energy source for green plants and is usually considered to be used very inefficiently because rather less than 1 % of the total radiation received per unit area of land actually appears in the product.

This ignores a number of important features. Somewhat under half of the radiation that falls on plants is actually usable, since it has to be in the $0.4–0.7\,\mu m$ wave region of the spectrum. Much of the incident light is reflected, sometimes on to another leaf, but if it were not so, we should not be able to see the plants. In other words, sunlight is doing many other essential things besides providing energy for plant growth, including maintaining the whole environment at a temperature that allows growth to take place. Even for plant growth, the amount used in photosynthesis is small relative to that used by plants in evapo-transpiration (Table 11.8).

TABLE 11.8

ENERGY USED BY PLANTS IN PHOTOSYNTHESIS AND EVAPO-TRANSPIRATION (A RANGE OF VALUES MEASURED FOR DIFFERENT CROPS IN DIFFERENT PARTS OF THE WORLD)

Gross daily insolation $(KJ\,cm^{-2})$ (A)	Percent of A used in photosynthesis	Percent of A used in transpiration
2·5–2·9	1·2–2·8	34–58

This enormous evaporation of water from warmed leaves is absolutely essential to plant growth and the energy used in this way is just as much a part of the real energy cost of production as is that employed in photosynthesis. It may be argued that this emphasises the inefficiency still further but, if this did not happen, the hydrological cycle would still have to be powered and the environment as a whole kept warm enough for plants, animals and men to tolerate.

So plants use rather more of the solar radiation they receive than at first appears and only a small proportion of the total is completely available for photosynthesis.

Even so, it has been calculated that at normal light intensities the theoretical maximum conversion of available visible radiation is of the order of 5–6 %, at times when temperature is adequate and water and nutrients are non-limiting.

Much therefore depends upon water and nutrient supplies and both may be supplemented. Irrigation is quite widespread throughout the world, although it has been estimated that less than 100 million hectares have been irrigated out of a potentially irrigable world total of 200 million hectares.

World fertiliser use is substantial, amounting to 70·5 million tonnes of N, 34 million tonnes of P_2O_5 and 25·9 million tonnes of K_2O in 1984/85. All these inputs involve substantial costs, including support energy, and a greater use of nitrogen-fixing legumes would seem to be one sensible answer.

CROP PRODUCTION PROCESSES

It has been estimated (Duckham *et al.*, 1976) that only 2·8 % of the possible world potential net photosynthate is actually formed in world farming, because less than half of the world's cultivable land is cultivated, only 70 % of this bears a crop in any one year and only 70 % of the growing season is actually used.

About 22 % of the potential net photosynthate is therefore used in the production of human food and the difference between this figure and the final 2·8 % is reckoned to be due to losses, bad husbandry, weather uncertainty and inadequate crop canopies. There are further losses in harvesting, processing, storage, distribution and within the consuming household.

There is every reason, therefore, to try and form a picture of the crop production process and to identify where losses and inefficiencies occur.

The main process can be visualised as an energy flow diagram (Fig. 11.2),

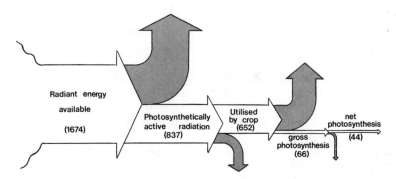

FIG. 11.2. Energy flow in crop production. The numbers in brackets are 10^8 J/ha/day: losses are shown shaded. The proportion of the total incident radiation appearing as chemical energy in the crop = 2·7 %.

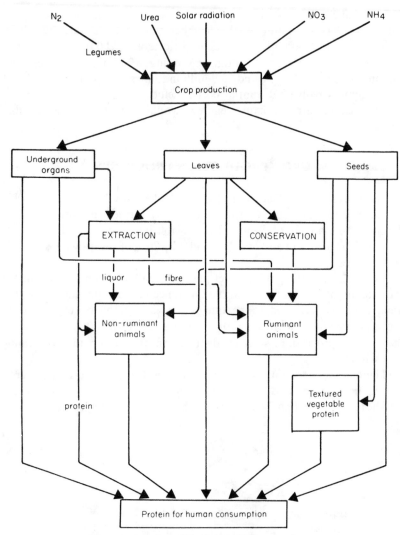

FIG. 11.3.   Protein production from crops.

bearing in mind all the management and input variables referred to in Chapter 2.

Protein production can best be described somewhat differently (Fig. 11.3) in order to take into account the possibility of green crop fractionation (Pirie, 1971, 1987) as a method of extracting the protein in

TABLE 11.9

CHARACTERISTICS OF THE MAJOR WORLD CROP PRODUCTION SYSTEMS

(Based on Cooper, 1976)

	Dry matter (t/ha)	Percent fixation of photosynthetically active radiation
Forage Crops		
Perennial ryegrass (C₃)*	25	3·0
Sorghum (C₄)	47	2·6
Napier (C₄)	85	4·9
Plantation Crops		
Sugar cane (C₄) sugar	22	1·2
Oil Palm (C₃) fruit	11	1·1
Cereals		
Wheat (C₃) grain	6	1·1
Barley (C₃) grain	6	1·1
Rice (C₃) grain	12	1·3
Maize (C₄) grain	13	1·5
Roots and Tubers		
Potato (C₃) tubers	12	1·6
Sugar beet (C₃) sugar	8	1·0
Cassava (C₃) tubers	22	1·0

* C_3 and C_4 plants have characteristically different respiration pathways.

leafy crops (including grass) that is surplus to the needs of ruminant animals. This kind of partitioning will be further considered in Chapter 13.

The processes illustrated in the figures are to be found in most crop production systems, although some of the pathways may be omitted and some additional cycles and feedbacks may occur.

The attributes that differentiate the main systems are (a) the type of crop plant (annual or perennial, tree or root, etc.) and its product; (b) the nature of the rotation (fallow, ley); (c) the level of inputs (labour, capital, machinery) and (d) the scale of the operation.

The major world crop production systems and some of their characteristics are listed in Table 11.9.

In general, crop *production* systems are visualised as finishing 'at the farm gate', the products being such commodities as grain, potatoes or raw cabbages, but, in fact, we do not eat such products without considerable further processing and preparation, including cooking. These processes involve their own additional costs and usually some losses of material. On

TABLE 11.10
AGRICULTURE'S PRODUCTION OF RAW MATERIALS FOR THE FOOD (AND BEVERAGE)
INDUSTRY
(From Spedding, 1985)

Farm gate product	% Processed[a]	Examples of product
Raw cow's milk	97[b]	Pasteurised or sterilised liquid milk, yoghurt, cheese and butter
Sugar beet roots	100	Sugar, molasses and pulp
Cereals	100	Flour, breakfast food and malt
Live animals for slaughter	100	Carcase meat, cooked, cured and smoked meats, pies
Raw potatoes	51[b] (The remainder will be cooked before consumption)	Canned, crisped dehydrated and frozen

[a] In addition to transporting and marketing.
[b] 1979/80.

the other hand, the purpose of processing is usually to render foods more nutritionally valuable, more attractive to eat, easier to store or even to eliminate toxic principles.

All these additional processes may alter some aspects of the efficiency of production but it is difficult to take them into account because they are so variable. You *can* eat raw cabbage and there are many different ways of cooking it, so the cost of cooking cannot be added on as an unavoidable, constant charge.

Similarly, there are marketing and distribution costs that are often unavoidable but these, too, are extremely variable and difficult to associate, quantitatively, with particular crop products.

In developed countries, a very high proportion of the food consumed is processed by a large and complex food industry (see Table 11.10). Where this is so, marketing may mean many different things. Farmers may have to market their produce to industry: industry then markets a great variety of products based on what they buy from farmers, but mixed, processed, preserved, treated and added to, packaged and presented.

In developing countries, much more of the food processing and preparation is done in the consumer household and marketing only applies to what is sold.

All this applies to both crop and animal products.

12

Animal Production Systems

Remarkably few animal species are used agriculturally, although many more, such as pests and disease organisms, are of agricultural importance. Those that give rise to products of major importance are listed in Table 12.1. Many of them also produce valuable by-products such as offals, bone meal, dung (used for fuel and the construction of huts), hair, feathers and glue; and some of them are also used for transport and traction. In many parts of the world, animals are still the main source of power for irrigation, cultivation, milling and transport of people and materials. Animals are also employed for a great variety of purposes, such as guarding and herding, that are less directly connected with a production process.

Most of the major animal production systems are based on warm-blooded animals (homeotherms). Of the cold-blooded animals (poikilotherms), fish and shellfish have been farmed for hundreds of years, as have two insects (the bee and the silk moth); crustaceans and snails are amongst those currently being explored. A list of the less important animals (i.e. on a world basis; they may be very important locally) is given in Table 12.2.

Table 12.1 also gives the estimated numbers of the major farm animals of the world but, since one mouse does not equal one elephant, the size or weight of the animals also has to be taken into account.

THE ROLE OF ANIMALS

As has been implied already, farm animals play several roles within agriculture, the most important being:

(1) The production of useful products.
(2) The collection of dispersed nutrients.

TABLE 12.1
THE MAJOR AGRICULTURAL ANIMALS
(Source: FAO, 1985a)

	Species	World population 1984 (thousands)
Cattle	Bos taurus Bos indicus	1 272 541
Buffalo	Bubalus bubalis	126 102
Musk oxen	Ovibos moschatus	N/A
Reindeer	Rangifer tarandus tarandus	N/A
Sheep	Ovis aries	1 139 520
Goat	Capra hircus	459 575
Horse	Equus caballus	63 871
Ass	Equus asinus	39 866
Mule		15 279
Camel	Camelus dromedarius Camelus bactrianus	17 207
Llama	Lama glama	N/A
Alpaca	Lama pacus	N/A
Pig	Sus scrofa	786 668
Hen	Gallus domesticus	7 305 000
Duck	Anas domesticus	159 000
Turkey	Meleagris domesticus	161 000
Goose	Anser domesticus	N/A

N/A = not available.

(3)   The conversion and concentration of nutrients.
(4)   The performance of work.

The first three are concerned with the production of food, although the first includes all other products as well: the animals mainly involved in the performance of work are given in Table 12.3.

The collection of dispersed nutrients is well exemplified by the ruminants, such as cattle and sheep. These are able to digest fibrous feeds (grasses, shrubs) that Man cannot use, and are thus able to utilise land that grows herbage but cannot be economically cropped. It is therefore of great importance to understand how animals are able to digest feeds that are of no direct value to Man. The ruminants depend upon microbial populations located in the enlarged first part of the stomach (the rumen); horses and rabbits also depend on microbes but these are located in the colon and caecum, respectively. Snails secrete their own cellulase enzymes but

TABLE 12.2
SOME LESS IMPORTANT AGRICULTURAL ANIMALS

Mammals	Yak	*Poephagus grunniens*
	Eland	*Taurotragus oryx*
	Red deer	*Cervus elephas*
	Rabbit	*Oryctolagus cuniculus*
	Guinea-pig	*Cavia porcellus*
	Capybara	*Hydrochoerus hydrochoeris*
	Mink	*Mustela vison*
	Dog	*Canis familiaris*
Birds	Guinea fowl	*Numida meleagris*
	Quail	*Coturnix coturnix*
Fish	Silver carp	*Hypophthalmichthys molitrix*
	Grass carp	*Ctenopharyngodon idella*
	Tilapia	*Tilapia* spp.
	Milkfish	*Chanos chanos*
	Catfish	*Pangasius* spp.
	Trout	*Salmo* spp.
	Carp	*Cirrhinus* spp.
Invertebrates	Clam	*Mercenaria mercenaria*
	Mussel	*Mytilus* spp.
	Oyster	*Ostrea* spp.
	Snail	*Helix pomatia*
	Bee	*Apis mellifera*
	Silk moth	*Bombyx mori*

TABLE 12.3
ANIMALS USED FOR WORK

		Nature of work	
Animal	*Species*	*Traction*	*Transport*
Oxen	*Bos taurus*	√	√
	Bos indicus	√	√
Buffalo	*Bubalus bubalis*	√	√
Yak	*Bos grunniens*		√
Reindeer	*Rangifer tarandus*	√	√
Horse	*Equus caballus*	√	√
Ass	*Asinus* spp.	√	√
Mule	*E. asinus* × *E. caballus*	√	√
Camel	*Camelus dromedarius*	√	√
Llama	*Lama glama*		√
Dog	*Canis familiaris*		√

herbivorous fish, such as the Chinese Grass Carp (*Ctenopharyngodon idella*) have no special adaptations and simply consume a great deal of herbage, which passes rapidly through the alimentary tract, during which cell contents are digested.

Simple-stomached animals (such as pigs and poultry), that need the same sort of food as we do ourselves, may compete with humans or live on waste materials and by-products. In terms of utilising household waste, it is a positive advantage to have animals that can live on similar foods to the ones we use (and waste).

The conversion of nutrients by animals may be from a form that we cannot use or from a form that we do not wish to use (e.g. fish heads), into animal products that we value. In most cases, however, only parts of the animal are valued for food and a proportion is wasted or used for other purposes (see Table 12.4 for an example).

One effect of animal production is to concentrate the nutrients that were present in the plant material: this has advantages for us, in terms of the

TABLE 12.4
PRODUCTS AND BY-PRODUCTS FROM A BEEF ANIMAL
(Based on *Farmers Weekly*, 1976)

Component	Percent of liveweight (c. 454 kg)	Derived products
Carcass	55·0	Meat, bone meal, tallow
Intestines (empty)	2·2	Casings, surgical ligatures
Liver and gallbladder	1·5	Dyes, gallstones
Pancreas and spleen	0·3	Insulin, chymotrypsin
Hide	7·0	Leather
Feet	2·0	Buttons, gelatin
Head	3·2	Bone meal
Blood	3·5	Glues
Stomachs (empty)	2·5	
Intestinal fat	2·5	
Caul fat	3·2	
Heart and lungs	2·0	
Skirt	0·3 ⎫	Soap, chemicals, pet food, fertiliser,
Tail	0·3 ⎬	plastics, paint, textiles, lubricants,
Gut contents	14·5 ⎭	glycerine, cosmetics, synthetic rubber[a]
	100·0	

[a] These are derived from fats, bone and offals and thus from parts of the carcass as well.

palatability and nutritive value of what we eat, and also in terms of the costs of transport, handling and storage of the product.

The performance of work by animals may be an important part of an animal production system and has the advantage of using solar radiation, indirectly, rather than fossil fuels. In many cases, animals are dual- or multi-purpose and may produce food as well as perform work.

THE NEED FOR ANIMAL PRODUCTS

Man does not need animals for food, in the strict sense that he could manage without them. The world would be a less interesting place without animals and we would be deprived of many products that we enjoy: nonetheless, animals are no more essential to us than other species are to a cow.

TABLE 12.5

AN INDICATION OF THE NUMBER OF PEOPLE WHO COULD BE SUPPLIED WITH ADEQUATE ENERGY OR PROTEIN PER HECTARE (PER YEAR) FROM DIFFERENT PRODUCTION SYSTEMS

(Clearly the actual values depend upon the precise needs of people and the level of production per hectare.)

Production system	Energy[a]	Protein[a]
Soyabeans	5	14
Maize	10·4	5·2
Rice	14	7
Wheat	8·4	6·3
Potato	16·5	9·5
Sugar beet	34	12
Milk	2·5	3
Poultry meat	1	2·5
Pig meat	2	1·4
Beef	1	1
Lamb	1	1

[a] Gross values do not take into account digestibility or biological value: the number of people supported therefore varies with the weightings attached for such attributes, but considerable allowance has been made in these calculations.

TABLE 12.6
VEGETARIAN DIETS

An Israeli[a] example Daily consumption	(g/head)	Percent of total dietary calories provided by items in average diets[b] of Vegetarians and Vegans[c]	
Bread	235	Milk	15	—
Pulses	39	Cheese	2	—
Nuts and seeds	103	Eggs	2	—
Oils and fats	34	Cereals	15	16
Fruits and vegetables	1 718	Potatoes	6	9
Sugar and sweets	31	Pulses ⎫	20	20
		Vegetables ⎭		4
		Fruit ⎫	16	15
		Nuts ⎭		12
		Sugar, sweets and beverages	11	10
		Butter or margarine	11	13

[a] Guggenheim et al. (1962).
[b] Wokes (1956).
[c] Vegan diets contain no food of animal origin.

However, this oversimplifies several issues. For example, Eskimos could not inhabit frozen lands if there were no fish, seals or reindeer, and people could not inhabit and live off the great permanent grasslands of the world if these were not converted to usable human food by ruminants. So we could only survive as vegetarians in certain areas if we imported our food and, in many cases, this would be both impracticable and uneconomic.

Similarly, peasants in Africa, India and Asia could not cultivate enough land to feed themselves without the use of cattle and buffalo and the freedom of the rest of us from this need depends primarily on oil.

When it comes to the supply of human food from croppable land, however, it is clear that more people can be supported per hectare if they feed on crop products rather than on animal products (Table 12.5).

Furthermore, there is no evidence that Man cannot live satisfactorily on crop products alone, provided that he eats a wide enough range to satisfy all his dietary needs, especially for essential amino-acids. (Table 12.6 illustrates such a diet). It may be difficult, however, to obtain sufficient vitamin B12 from crop products and the quantities needed to supply adequate protein may be excessive (see Table 12.7).

In fact, of course, it is nearly always possible to supplement a vegetarian diet with a variety of animal products (caught, hunted or farmed) without major animal production systems.

TABLE 12.7

AMOUNTS OF VARIOUS FOODS REQUIRED TO PROVIDE THE DAILY PROTEIN RE-QUIREMENT (70 g) OF A 70 kg MAN

Food	Amount (g)
Cassava	4 600
Milk	2 000
Bread	940
Rice	880
Egg (without shell)	434
Meat	400
Beans	328
Poultry meat	234
Dried fish	117

Where such systems are practised, the main reasons are usually that there is a demand for the products because people like them, want them and are willing and able to pay for them. Sometimes, as in the case of some non-food products, such as wool, it may be because there is no substitute or no satisfactory substitute. It is probably true that eventually a satisfactory substitute will be found for all natural products but it is not necessarily true that such substitutes can be produced at a competitive cost.

Increasingly, the efficiency with which scarce resources are used in production systems will govern which systems survive and the precise form they take.

EFFICIENCY IN ANIMAL PRODUCTION SYSTEMS

This is a large subject and can only be dealt with briefly here.

If we are thinking chiefly of food products, then it is clear that the way the output is expressed in an efficiency ratio will depend upon whether we are considering feeding people or satisfying preferences.

Thus, if lamb is wanted, there is no way in which beef, milk or anything else can produce more efficiently. On the other hand, if we wish to satisfy nutritional needs, we can compare different animals in terms of energy, protein, lysine, iron or any other valued constituent.

As far as resources are concerned, any one could be critical in given circumstances (such as irrigation water) but land, labour and capital are

TABLE 12.8
EFFICIENCY OF RESOURCE USE

Product	Quantity of resource required to produce 1 MJ of gross energy in the product			
	Land (ha)	Solar radiation (MJ)	Labour[a] (min)	Money[b] (p)
Milk	0·000 114	3 762	0·185–0·266	1·85
Eggs	0·000 2	6 600	0·18–0·68	6·3
Lamb	0·000 33	10 890	0·45–0·65	4·0
Beef	0·000 33	10 890	0·23–0·29	3·5
Poultry meat	0·000 22	7 260	0·34–0·52	0·27

[a] All costs, including fixed and working capital (see Spedding and Hoxey, 1975).
[b] Based on Nix (1976).

usually important (Table 12.8). A useful resource to use as an illustration of the factors influencing efficiency in animal production is feed, since this is usually a crop product, subject to the influences on efficiency of land use dealt with in Chapter 11. Feed is thus the intermediate output of land, on which animal production is based.

Feed conversion efficiency is the usual measure of feed use by animals. It is often expressed as a ratio of feed used per unit of animal product but it is generally better to use an efficiency ratio, of product output per unit of feed. This is illustrated for different animal production systems in Table 12.9. Outputs per unit of land are also given, since not all crop yields are the same: even so, it has to be remembered that not all these feeds can be produced on the same kind of land or in the same environment.

Efficiency of feed conversion can be directly influenced by many components of the environment, including climate and disease, and by a number of characteristics of the animal populations used.

Obviously the species, breed and strain may be important, and the level of nutrition must matter, since no production will occur at all if nutrition is inadequate.

In addition, management may be of over-riding importance. For example, if the bull or the ram is not made available to female animals at the right time, no amount of nutrition is going to result in progeny.

The most general expression of all these factors is the performance of the individual animal or population.

TABLE 12.9

FEED CONVERSION EFFICIENCY (FCE) AND OUTPUT PER UNIT OF
LAND FOR SELECTED ANIMAL PRODUCTION SYSTEMS
(Typical values from the literature)

Production system	FCE^a	Energy output $(MJ/ha/y)$
Milk	12	10 450
Eggs .	11	4 807
Chicken	14	4 600
Beef cattle (edible meat)	6	3 100
Suckler cow + calf (carcass)	3	3 924
Ewe + lamb(s) (edible meat)	3	2 100–5 400
Pig (edible meat)	24	7 900
Rabbit (carcass)	8	7 400

$$^a\ FCE = \frac{\text{Total energy in product}}{\text{Gross energy in feed for the progeny and a}} \times 100$$
proportion of the parents' feed

Since warm-blooded animals require a certain amount of food for 'maintenance' purposes, this can be regarded as a kind of 'overhead' cost that has to be spread over the volume of production. It follows that the greater the production, the smaller the share of the overhead cost that has to be borne by each unit of production. Thus the efficiency of output per unit of feed input rises with increasing levels of production.

The simplest example is that of an individual (sheep, for instance): with increasing quantities of feed consumed (F), liveweight gain (G) per day increases and so does the efficiency of feed conversion (E), as shown in Fig. 12.1.

The same kind of relationship holds for milk and egg production from individuals (Fig. 12.2). However, breeding populations always have to be maintained in order to sustain such individual production and this involves feeding both parental stock and those animals that are destined to replace them in due course. This means that the efficiency of feed conversion by populations is less than that of the productive individuals within them (as described in Chapter 4).

The calculation of such efficiency ratios may relate to different periods of time (a season, a reproductive cycle, a year, a lifetime) and may include some or all of the outputs (e.g. milk plus culled, old or uneconomic cows plus calves) and one or more inputs. The actual expression may also take many forms such as energy, protein or money.

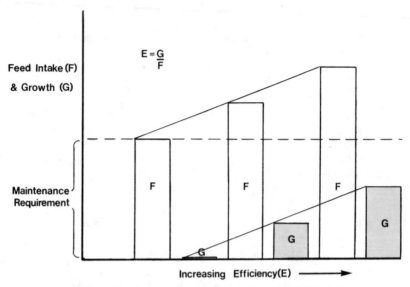

Fɪɢ. 12.1.   Feed conversion efficiency for an individual sheep.

Comparisons of the efficiency of production by different animal production systems have to be undertaken carefully, therefore, for clearly defined purposes. Very often the choice of system may be related to the availability of a particular resource: the question posed is then 'which animal production systems could use this resource efficiently?'. But it is then necessary to consider the implications in other directions and the effect on the efficiency with which other important resources are used.

ANIMAL PRODUCTION SYSTEMS OF THE WORLD

Animal production takes many forms, depending upon the animal species, the environment, the availability of resources and the product required. But the same product (e.g. milk) can be produced in a variety of different ways and even milk from cows or, more specifically still, from Friesian cows, can be produced in a whole range of different systems.

The major forms of animal production system found in the world are listed in Table 12.10.

In most cases there is a major product but the importance of multiple-purpose production systems should not be underestimated and the importance of by-products is also considerable.

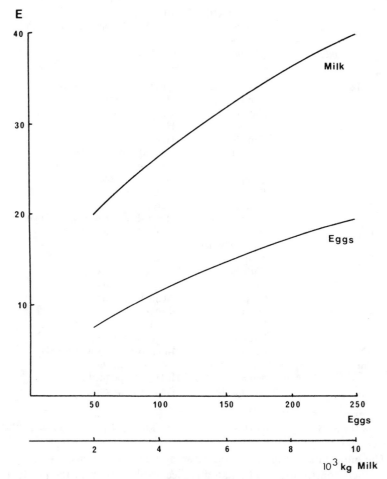

FIG. 12.2.   Efficiency of feed conversion in the production of milk and hen eggs.

$$\text{Efficiency } (E) = \frac{\text{total N output in products}}{\text{total N consumed by the animal population}} \times 100$$

The end product of these systems is usually thought of as what is sold by the producer but, in fact, there is often a great deal of processing beyond this point, quite apart from all the marketing and distribution activity that takes place after the original production process has finished.

It is also necessary to remember the substantial activities that produce the fertilisers, machinery, fuel, pesticides and other inputs to agriculture, as

TABLE 12.10

THE MAJOR ANIMAL PRODUCTION SYSTEMS OF THE WORLD

Product	Animals used	Main features
Milk	Dairy cattle	Grazed, housed, feedlot
	Buffalo	Housed, herded
	Sheep	Grazed, herded
	Goats	Herded
	Camels	Herded
Meat	Beef cattle	Grazed, housed, feedlot
	Buffalo	Range-grazed, housed, herded
	Sheep	Grazed, herded
	Pigs	Housed, penned
	Goats	Herded
	Poultry	Housed, penned or free-range
Eggs	Poultry (mainly hens)	Housed, penned or free-range
Fish	Tilapia	Lakes
	Catfish	Artificial ponds
	Trout	Artificial ponds
	Carp	Artificial ponds
Wool, hair and skins	Sheep	Range grazing
	Goats	Herded
	Rabbits	Intensive housing
	Crocodiles	Intensive, tanks and ponds
Fur	Chinchilla	Intensive, penned
	Mink	Intensive, penned
Silk	Silk moths	Intensive, in cages

well as the servicing industries providing all kinds of goods, services and advice.

The importance of animal production systems cannot easily be judged from their contribution to human diet (see Table 3.2) or by the products that enter into trade.

For one thing, there are no reliable estimates of animal production from smallholders and peasants and little is known of the extent and productivity of small-scale rabbit and guinea-pig keeping, for example. Even if their numbers were known, reproductive rates are so high and production cycles so short that it would be very difficult to estimate output.

Furthermore, many animals are kept for religious and cultural festivals and thus have a significance outside their direct contribution to production. Animals also represent an important form of wealth to many people, an insurance that can be called upon in case of need, such as ill-health, and whilst serving these purposes may also provide transport, draught power, milk, offspring, faeces for fuel or construction materials, and sometimes wool or blood as well.

Attitudes to animal production have been changing in recent years. In general, the proportion of animal products in the diet increases with increasing affluence, but recently, in developed countries, consumption of animal products has been viewed rather differently.

ATTITUDES TO ANIMAL PRODUCTION

A small minority of people believe that animals have rights and that they should not be exploited in any way. Others are simply concerned that, where animals *are* kept, they should be given certain necessary freedoms— to move about, stretch, turn around, and so on.

Animal Welfare has become a significant force in developed countries, especially where overproduction and affluence combine to negate the argument for increased production of the cheapest food possible.

It is, of course, a proper concern and it is understandable that it is generally directed against practices and systems (e.g. battery hens, tethered sows and crated veal calves) that are considered unacceptable. It is not so easy to be sure that alternatives are better: free-range animals can easily suffer from adverse weather, parasites and disease, poor nutrition and the activities of other animals.

There is therefore a strong onus on research workers to develop alternatives that are better for the animal, better for farm workers, productive and profitable. This also requires a systems approach, since any one change may have unforeseen consequences on other parts of the system. The question 'What is a better system?' is thus similar to that posed in Chapter 2, 'What is a better cow?'.

13

Industrial Food Production Systems

There are a number of food production systems that use little or no land (except for buildings) but do use biological organisms, usually in highly controlled environments. No-one would class them as farming, which seems generally to be associated with the use of land or water, but they are regarded here as part of agriculture.

It is hard to draw the line between what is industrial and what is not. Intensive piggeries and battery hen units are considered by many to be virtually industrial and it is difficult to argue that they represent farming but that single cell protein production from field beans does not.

Accepting these difficulties, it is nevertheless convenient to group together those highly controlled biological processes in which the individual organism is usually of small significance and land is not directly involved.

In recent years a trend towards such systems has developed in industrialised countries but the recognition that oil is becoming an increasingly costly and scarce resource has substantially reduced their future prospects. Some of them actually use oil as a feedstock and it can be argued that this is a better way to use fossil fuels than burning them.

This does not influence the cost of oil, however, and food derived in this way still has to compete economically with that produced in other food production systems.

The main industrialised food production systems may be exemplified by: (a) single cell protein; (b) prawn production; (c) meat analogue processes and (d) green crop fractionation.

154

TABLE 13.1

THE PROTEIN DOUBLING TIME (R) FOR
MICRO-ORGANISMS AND BEEF CATTLE
(After Worgan, 1973)

Organism	R (h)
Candida utilis	3–5
Fusarium semitectum	5
Escherichia coli	5
Beef cattle	2 800

TABLE 13.2

THE MAIN PLANT SPECIES USED FOR SINGLE CELL
PROTEIN PRODUCTION
(Sources: Heydeman, 1973; Worgan, 1973;
Walker, 1973)

Algae	Chlorella spp.
	Spirulina maxima
	Wolffia arrhiza
	Scenedesmus sp.
Bacteria	Bacillus megaterium
	Bacillus subtilis
	Escherichia coli
	Chlamydomonas spp.
	Brevibacterium flavum
Yeasts	[a]Candida utilis
	Saccharomyces fragilis
	Sacharomyces cerevisiae
	Endomycopsis fibuliger
Fungi	Asperigillus niger
	Neurospora crassa
	Fusarium semitectum
	Rhizopus oryzae

[a] This is the only microbial source of protein
currently produced in large quantities for use as
human food.

(a) SINGLE CELL PROTEIN

Unicellular organisms can increase in number at an astonishing rate, provided that they are supplied with their nutrient needs and an appropriate environment. Their potential rate of production per unit time is very high (Table 13.1), therefore, and it is not surprising that their usefulness as a food source is being explored.

The main processes being examined or developed are given in Table 13.2. Those that use crops, such as beans, as a feedstock, have to justify themselves by comparison with simply eating the beans. Those based on oil

TABLE 13.3

ENERGY USE IN INDUSTRIAL FOOD PRODUCTION SYSTEMS
(After Worgan, 1976)

Raw material used	Method of conversion	Energy input per unit of food energy produced
CO_2 and water	Methanol synthesis followed by yeast or bacterial conversion	14·9
Fossil fuel	Methanol converted by yeast or bacteria	5·6
CO_2 and water	Electrolysis of water and bacterial conversion	13·6
Cellulosic agricultural by-products	Chemical conversion to sugar followed by bacterial, yeast or fungal conversion	2·7
Cellulosic agricultural by-products	Chemical modification followed by bacterial or fungal conversion	<2·7

either have to argue that it is better to use oil for food production than to use it for other purposes or that the oil used is a waste product of little use as a fuel.

Those that do not involve photosynthesis are bound to use considerable quantities of support energy and it is this feature that renders them less promising than they once appeared.

The efficiency with which such processes use resources is illustrated in Table 13.3.

(b) PRAWN PRODUCTION

The use of cold-blooded animals has a number of theoretical advantages, especially in relation to their low maintenance requirements and high reproductive rates.

The prawn is used here as an illustration of one such animal that is being grown in highly controlled conditions to produce a product with a high monetary value. It would only contribute to the solution of a problem primarily concerned with feeding people if the prawns were fed on animal or vegetable wastes, but this is not currently the case. Several species can be

TABLE 13.4

PRAWN PRODUCTION

Main species used	*Penaeus japonicus*
	Penaeus monodon
Main countries engaged	Japan, China, Phillipines, Indonesia
Estimated output levels	2 000 kg/ha/y
Target output	5 000 kg/ha/y
Method of culture	Extensive ponds with water derived from rivers or estuaries

used (see Table 13.4) and their productivity can be high. Of course, with inadequate control losses can also be high.

The main disadvantages are the capital cost of tanks and the energy cost of warming water, at least in cooler climates.

(c) MEAT ANALOGUES

The object of producing meat analogues is to arrive at a product that looks and tastes like meat but costs much less because it is based on textured vegetable protein.

It is a more elaborate way of treating crop products and this additional processing costs money and energy: it produces no more nutrients than were in the original crop but it wastes less than processing through animals.

It may be hard to justify the extra costs and the lowered efficiency of support energy use (Table 13.5) unless meat-like qualities of texture and taste are regarded as very desirable. However, it might be possible to base these processes on otherwise unusable vegetation. This would bring them

TABLE 13.5

SUPPORT ENERGY USE IN INDUSTRIAL SYSTEMS OF FOOD PRODUCTION

(After Worgan, 1976)

System	Total energy input per unit of food energy output
CO_2 + water → methanol synthesis → yeast or bacterial conversion	14·9
CO_2 + water → electrolysis of water → bacterial conversion	13·6
Fossil fuel → conversion to methanol → yeast or bacterial conversion	5·6
Cellulosic by-products → chemical conversion to sugar → bacterial, yeast or fungal conversion	2·7

close to the idea of green crop fractionation—or what used to be called leaf protein extraction.

(d) GREEN CROP FRACTIONATION

This form of processing aims to extract a protein concentrate that can be used by Man or simple-stomached animals from vegetation that cannot be used directly except by herbivorous animals. In addition, there are fibrous residues and juice that can be fed to ruminants, for example.

Indeed, for the process to be economic it is necessary to envisage all the fractions being used.

The main advantage to be derived is that of a useful protein extract based on protein in the original material that was surplus to the requirements of ruminants. As a result of this, it is possible to obtain approximately the same levels of animal performance on the fibrous residue as was possible on the original vegetation (Table 13.6). In addition to this, the protein extract can be used to feed pigs or poultry or be further processed for direct human consumption. Because of the inefficiencies of conversion by pigs and poultry, much more human food is obtained if the protein extracted is consumed directly. There are, however, still some problems and extra costs associated with this.

In terms of resource use, green crop fractionation as currently operated makes good use of land and poor use of support energy. It can be done on a

TABLE 13.6

BEEF PRODUCTION PER HECTARE WITH AND WITHOUT GREEN CROP FRACTIONATION
(Derived from Spedding, 1976*b*)

	Beef (1) from pasture and silage	Beef (1) from dried grass	Beef (1) + LPC[a] (2) for humans	Beef (1) + LPC[a] (3) fed to poultry
Output of protein (kg/ha)				
(1) as beef	78	87	83	83
Additional output of protein (kg/ha)				
(2) as LPC[a] for humans	—	—	499	—
(3) as poultry meat produced from LPC[a]	—	—	—	50
Total protein output (kg/ha)	78	87	582	133

[a] Leaf Protein Concentrate.

small village scale, in which case capital investment is low, or on an industrial scale, in which case investment may be high.

The efficiency of land and support energy use is illustrated in Table 13.7 from which it can be seen that, if the protein extract can be used directly, it is possible to make efficient use of both. The biggest need for support energy is for drying and if solar energy could be used to power the process it would be able to make a big contribution to the feeding of people from crop by-products or otherwise unusable vegetation.

TABLE 13.7

EFFICIENCY OF USE OF SUPPORT ENERGY IN GREEN CROP FRACTIONATION
(Derived from Spedding, 1976*b*)

	Beef (grazing + conservation)	Beef + LPC for humans	Beef + LPC for poultry
Output of protein (g) as human food / Support energy used (MJ)	2·2	3·5	0·76
Energy requirement MJ/kg of protein	416	282	1 308

FUTURE DEVELOPMENTS

Green crop fractionation has a major advantage in that whole crops—at least above ground—can be harvested and all their constituents used. As Pirie (1987) has pointed out, green leaves represent the most abundant source of edible protein but the human digestive system has a limited capacity to deal with whole leaves.

Perhaps its major disadvantage is that vegetation has to be harvested and there are vast areas of rather sparse grassland where this would not be economic. The grazing animal serves as the harvesting agent in these areas.

However, quite apart from animal welfare considerations, some people believe that it is healthier to avoid animal products in the diet. If such an attitude should increase greatly, it would have a number of important effects. It would reduce the area required to feed people but exclude those areas that could only be harvested by grazing animals. It might also greatly increase the attractiveness of meat analogues and it must be assumed that industry will become increasingly skilled at using plant material as the basis of food products that have whatever properties the consumer demands.

The relative efficiencies of different sorts of production depend heavily on the demand for the products.

14

The Relative Efficiency of Production Systems

Comparisons of the efficiency of different systems can only be made where both the products produced and the resources used are sufficiently similar or where they can be expressed in comparable terms.

Economic comparisons are the most universally applicable since all products and resources can be expressed in monetary terms and these can be interpreted internationally. Such comparisons rank systems in order of profitability, for example, or return on capital for the particular costs and prices used in the calculation. Since the latter do not remain constant, this may be a poor guide to the future: it may be an essential guide to the farmer in the short term, however.

Similar considerations may govern the outlook of a nation, particularly if agricultural products form a significant proportion of its exports.

If we are concerned with agriculture in terms of the efficiency with which resources are used in feeding people it makes sense to compare systems as energy or protein producers or to assess the number of people they can feed. Even then, comparisons have to be based on the use of common resources in common environments. Since there are many different ways of looking at systems (see Chapter 3), there are bound to be at least as many ways of comparing them. A few of the most useful will be considered here.

CROPS VERSUS ANIMALS

As will have been clear from Chapters 10 and 11, there is a range of efficiency to be found *within* any production system, so it is important to base comparisons between them on comparable versions of each. Similarly, there is a wide range of efficiencies within crop production systems and within animal production systems. Nevertheless, crop production is generally more efficient than animal production in the output of dietary protein and energy on land that will grow crops (Table 14.1). This is so for the use of land area, incident solar radiation, water, fertiliser and support energy.

On land that cannot grow crops, or from which crops could not economically be harvested for direct human consumption, it may still be

TABLE 14.1

OUTPUT OF PROTEIN AND ENERGY PER UNIT OF LAND

(From Spedding and Hoxey, 1975)

Product harvested	Protein (kg/ha/y)	Energy (MJ/ha/y)
Crop		
Dried grass[a]	200–2 200	92 000–218 000
LPC	2 000	—
Cabbage	1 100	33 500
Maize	430	83 700
Wheat	350	58 600
Rice	320	87 900
Potato	420	100 400
Animal		
Rabbit	180	7 400
Chicken	92	4 600
Lamb	23–62	2 100–7 500
Beef	27–57	3 100–4 600
Pig meat	50	7 900
Milk	115	10 460

[a] Not directly consumable by Man.

possible to use animals. On such land, animal production is, of course, vastly more efficient in the use of all resources.

Since many of the farming systems described in Chapters 8 to 13 are practised on quite different kinds of land and in different climates, it is not always possible to make direct comparisons between them. Within these constraints, however, it is useful to examine their productivities per unit of land, solar radiation and support energy (Table 14.2) because this indicates the size of human population that can be associated with these systems in the areas where they are found.

TABLE 14.2

RELATIVE EFFICIENCY (E) OF ENERGY AND PROTEIN PRODUCTION IN AGRICULTURAL SYSTEMS

(Source: Spedding and Walsingham, 1975)

Production system	E^a for the use of			
	Land[b] (ha)	Solar radiation[c] (MJ)	Support energy[d] (MJ)	Fertiliser N (kg)
Milk[e]: Energy	15 576	0·000 47	0·54	129
Protein	210 × 10³	0·006 4	7·30	1 736
Beef[f]: Energy	5 772	0·000 17	0·11	22
Protein	90 × 10³	0·002 7	1·66	339
Sheep meat[g]: Energy	4 929	0·000 15	0·23	38
Protein	53 × 10³	0·001 6	2·50	404
Wheat[h]: Energy	58 600	0·001 78	5·4	586
Protein	350 × 10³	0·010 6	32·0	3 500
Potatoes: Energy	100 400	0·003	5·1	619
Protein	420 × 10³	0·013	21·1	2 593

[a] E is expressed as energy (MJ) or protein (g) output of milk, carcasses, grain or tubers, per unit of resource used, on an annual basis.

[b] Assumed to be receiving nitrogenous fertiliser at the annual rate of (a) 121 kg/ha for milk, (b) 265 kg/ha for beef, (c) 131 kg/ha for sheep meat, (d) 98 kg/ha for wheat and (e) 125 kg/ha for potatoes.

[c] Based on annual radiation receipt of 33 × 10⁶ MJ/ha/y.

[d] Support energy is defined here as the additional energy (labour, fuel and electricity) used on the farm, plus the 'upstream' energy costs, i.e. those used to manufacture the major inputs (fertilisers, machinery, herbicides, etc.; human labour is excluded) and the 'downstream' energy costs of processing and distribution.

[e] Permanent pasture dairy farm, approximately 76 ha and 100 cows averaging 900 gallons (4 217 kg).

[f] Suckler herd, intensive grassland system.

[g] Lowland fat lamb system.

[h] Winter wheat.

SUPPORT ENERGY VERSUS LABOUR

The picture that emerges is the familiar one that 'developed' agricultural systems are more productive per unit of land and solar radiation but less efficient in their use of support energy. Since support energy may represent higher inputs, such as fertiliser, that raise output per unit of land, this is not surprising. But support energy may also represent machinery and fuel that displace human labour or animal power and the effect may be quite different in these two cases. Where human labour is displaced, output per unit of land may be no higher—it may even be lower. This rather depends upon how much labour is available, especially at critical times when tasks such as cultivation and harvesting have to be accomplished rapidly. Such 'timeliness of operations', in relation to crop physiology or the weather, may be easier to manage with machines, although individual selectivity may still benefit from human labour: this can be seen in the selection of appropriate leaves in the harvesting of tea by hand. Of course, the use of machinery will naturally lead to higher output per man, although it is hard to be sure that this is always so when the labour involved in making the machinery and supplying the fuel is all taken into account.

In the case of animal power, it seems fairly clear that net output per unit of land is lower but that output per unit of support energy is higher (Table 14.3) where animals are used instead of tractors, but it is hard to make sensible comparisons with human labour.

Animal production is always based on plant production of some kind and adds its own losses to those incurred in the initial crop phase. The conversion of crops that can be consumed directly can never avoid a decrease in biological efficiency, therefore. This can be well illustrated for the use of nitrogenous fertiliser, an input that is only required for the crop production part of the process. Although no more is added during the

TABLE 14.3

ENERGY BUDGETS FOR ANIMAL POWER VERSUS TRACTORS

Proportion of food energy[a] produced per hectare that is required to supply	
a tractor	a horse
c. 3 %	c. 7 %
Support energy required to produce and run a tractor on a cereal farm = 2·8 GJ/ha/y	Feed energy required per horse per year = 68 GJ/y
	Support energy required by the horse (harness, stable, shoeing, etc.) = c. 15 GJ/y
Gross energy produced by cereals = 96·8 GJ/ha/y	
	Gross energy produced per horse per year as cereals[b] = 1 162 GJ

[a] Obviously 'feed energy' could not, in fact, be used directly for the tractor: gross energy is used throughout the calculation.

[b] Assuming the same cereal output as in the 'tractor' column.

TABLE 14.4

EXAMPLES[a] OF NITROGEN RECOVERY IN CROPS AND ANIMALS

	Amount recovered in the plant or animal as a percentage of the nitrogen applied as fertiliser
Pangola grass	79–103
Ryegrass	75
Beef (individual cattle)	8
Milk (individual cattle)	9·5

Animal populations	Amount recovered in animal products as a percentage of that in feed (from calculations by Wilson, 1973)
Beef	6
Milk	24
Hen eggs	20
Broiler chickens	20
Rabbit	17
Lamb	4

[a] Taken from actual experiments reported in the literature.

animal production phase, the reduction in output is such that production per unit of fertiliser is greatly decreased (Table 14.4).

NEED AND DEMAND FOR FOOD

In economic terms, of course, the conversion of crop products into animal products often results in an increase in economic efficiency—or it would not be undertaken—due to the higher value placed upon a unit weight of animal product. This can be illustrated for a variety of foods by expressing value as the relative price paid per kilogramme of protein (Table 14.5). These values reflect demand and do not necessarily have anything to do with how many people can be fed.

In any case, many foods are not well balanced nutritionally and few people live on only a small number of foods. Most people in developed countries enjoy considerable variety and are unlikely to suffer specific deficiencies: there are estimated to be millions of people in the world as a whole, however, who are malnourished in one way or another.

TABLE 14.5
PRICE PER KILOGRAM OF PROTEIN FOR ANIMAL AND PLANT
FOODS

	Product	World price[a] (£)
Animal	Beef	10·08
	Poultry meat	3·28
	Milk (cond. & evap.)	6·87
	Cheese	6·44
Crop	Wheat	0·86
	Wheat flour	1·27
	Rice	3·26
	Soyabeans	0·47
	Potatoes	5·33
	Cocoa beans	5·70

[a] Based on FAO (1985b) world average export unit values.

If we consider the number of people who could be adequately nourished per hectare of any one product, the answer would be determined by the output of whatever was the limiting nutrient. In general, human dietary requirements (see Table 14.6) can be considered, from a quantitative point of view, in terms of energy and protein. The number of people who could be supplied with their energy needs from a hectare of various products was shown in Table 12.5, alongside the number whose protein needs could be met.

The discrepancies between these two sets of numbers reinforce the notion that some foods supply energy and some protein but, in fact, there are only

TABLE 14.6
DAILY DIETARY NEEDS OF MAN FOR PROTEIN AND ENERGY

Age (years)	Male (♂) or female (♀)		Body weight (kg)	Energy (MJ)	Protein (g)
0–1	♂	♀	7	3·3	20
7–9	♂	♀	25	8·8	53
15–18	♂		61	12·6	75
18–35	♂		65	11·3	68
18–55		♀	55	9·2	55
55+		♀	53	8·6	51
Late pregnancy		♀	55+	10·0	60
Lactation		♀	55+	11·3	68

TABLE 14.7
PROTEIN CONTENT OF FOOD
(Percent of fresh weight)

	At farm gate (%)	As consumed[a] (%)
Wheat	8·9–13·6 (DM)	7·9–12·8 (dry flour)
Rice	8	2·1 (polished and boiled)
Potatoes	1·9–2·1	1·4 (boiled)
Beans	22–28 (dry)	4·1 (boiled)
Cheese (Cheddar)	26	26
Milk	3·4	3·4
Dried skim	36	36
Fish	—	16–24
Dried		37
Beef	15–20	15·5–29·6[a]
Pork	9–20	17–28·5[a]
Eggs	11–12·9	11–12·9
Lamb	12[b]–14	15–26·6[a]
Chicken	11–19[b]	20·6–23·4

[a] The higher values tend to be cooked lean portions.
[b] Carcass.

a few of the major products that supply insignificant amounts of protein. More serious, perhaps, is any discrepancy between the protein content of products as sold from the farm and their content on the plate of the consumer (Table 14.7).

EFFICIENT USE OF RESOURCES

Losses that occur after harvesting represent a substantial area for the improvement of many aspects of efficiency, although the avoidance of loss frequently involves an additional expenditure of energy. The same is true of the utilisation of by-products and waste material. Nevertheless, it is likely that efficient use of all that is produced by photosynthesis will be a major aim of future agriculture, even to the extent that part of the output may be used as a source of fuel to power the essential processes without recourse to oil.

Green crop fractionation has been devised as a method of harvesting the maximum photosynthate and then partitioning it into separate fractions suitable for direct consumption by Man and conversion by animals (see Chapter 11). The relative efficiency of this process compared with total conversion of the same crop by beef cattle is shown in Table 13.6. The extraction of protein that is surplus to the requirements of cattle (or sheep) makes it possible to feed the protein extract to Man, pigs or poultry whilst still obtaining almost as much beef as if the original crop had been fed. Because of the very high total output of protein where direct consumption by Man is involved, the support energy required per unit of protein output is not high, but the dependence on support energy per hectare is considerable. If ways could be found to fuel the process by the direct or indirect use of solar radiation, the process would be very efficient indeed in the use of all the major resources.

The same might be said of the fully industrialised processes. They excel in production per unit of time and space but they require substrate and this either has to come from land receiving solar radiation or from fossil fuel. If the substrate could not be used directly or if the land could not produce as much for direct consumption from some other crop, then processing in a highly controlled fashion might be sensible. But it is unlikely to have a firm future if it depends too heavily on fossil fuels.

EFFICIENCY IN THE FOOD CHAIN

Since a high proportion of food is now processed in developed countries, it is important to consider efficiency over the whole food chain. An example will illustrate the reasons for this (see Table 14.8).

Wheat is more efficiently produced than milk, if the calculation is up to

TABLE 14.8
EFFICIENCY OF ENERGY USE IN THE FOOD CHAIN

	MJ of energy in product per MJ support energy used
Wheat at farm gate	3·2
Bread—white, sliced, wrapped	0·5
Milk at farm gate	0·65
Milk bottled and delivered	0·595

the farm gate, for reasons already given, for the use of support energy, as for other resources. But milk can be drunk virtually as it is and usually only incurs rather small additional energy costs in transport and treatment. Wheat, on the other hand, is normally ground and cooked, involving processes that incur very heavy energy costs, quite apart from the costs of storage, transport and packaging.

If the two products are compared at the point of consumption, therefore, the large difference evident at the farm gate completely disappears.

Thus conclusions may be different for calculations made up to the farm gate, within the food industry, or over the whole food chain.

15

Agriculture and the Community

Agricultural activities influence the environment and impinge on the community in a number of different ways. A very high proportion of the population of many countries, however, live in towns and may not be greatly affected by changes in the appearance and smell of the countryside. Indeed, one effect of the development of agricultural systems has been a great reduction in the number of people *directly* engaged in agriculture.

Thus, relatively few people in a country such as the UK have any direct contact with agriculture and even fewer have a firm grasp of what it is about.

Well, why should they have? Could not exactly the same be said of deep sea fishing, or ship-building, or cigarette manufacture?

One purpose of this chapter is to consider why agriculture may be different and to argue that an educated citizen ought to know what agriculture is about and why it matters.

Since the proposition is framed in relation to 'an educated citizen', much depends upon the meaning attached to this term.

By 'educated' I mean intellectually equipped with the information, understanding and methodology that is needed to 'do' something of value for the community. This covers all kinds of education for all kinds of purposes within specified communities. It is difficult to talk about an educated person without relating one's comments to their needs in contributing to a particular community. A man may be a very good plumber without being described as 'educated'. If he then turned out to know all the works of Beethoven, he could certainly be described as knowledgeable about that subject, but would he thereby be *educated*? He might be described as *musically* educated, perhaps, but this suggests an additional purpose in life rather than an educated plumber. What *is* an educated plumber?

It could be argued that he is one who can contribute his special skills in an intelligent manner: that is, that he also understands the structure of houses, the vagaries of the weather and the needs of people. It could also be argued that, besides being a plumber, he is also other things, such as a father, a husband and an uncle. There are probably many different subjects that could be said to contribute to the education of our plumber in each of these roles. Thus education has purposive elements and the question 'education for what?' is a legitimate one to pose. That is why I have deliberately chosen the phrase 'educated citizen', since the argument concerns men and women in their roles as citizens, whatever other roles (domestic or professional) they may also have.

What, then, is a citizen? The most useful definition of such a concept will surely include more than 'an inhabitant of a city' or 'a person holding some special status within the community'. There has to be some element of participating membership of an organised, identifiable community, involving both rights and responsibilities, but, above all else, in relation to the community (city, state or world) as a whole. This last point imposes a responsibility on the citizen to be able to take part in debates on those issues which are in some sense central to the well-being of the state. Issues such as pollution control, voting procedures, major transport systems and the law all fall within this category. The citizen cannot possibly be expert in all of these things but he should never be entirely dependent upon those who are, because specific expertise is always too narrow for the purpose of integrating one activity into that of the whole.

So it is with agriculture. I am arguing that agriculture should be included in the list of subjects of major concern to the citizen but that he does not have to be anything like an expert agriculturalist. After all, if anything like this concept of citizenship is to be feasible, then it must be possible to grasp the essentials of vital subjects without acquiring specific expertise or detailed knowledge.

Why should agriculture be included? There are three main reasons that immediately spring to mind. First, the importance of food, overwhelmingly an agricultural product and likely to remain so (although this prediction should not be accepted unquestioningly), and literally vital to everyone, virtually every day. This means that food production should never become the concern of only a few, since it is of such vital interest to all. Secondly, everyone has a genuine interest in the diet available to him, not only quantitatively but also with regard to its quality, cost and variety. This means that everyone ought to have some say in what is produced (since not everything can easily or cheaply be grown everywhere). Thirdly, as world citizens, we should be concerned about the hungry people of the world and how our activities affect *their* food supplies and their capacity to pay for them.

If the citizen does not share these concerns, he hardly fits the earlier description: if he does, then his ability to influence matters depends upon his being adequately educated for this purpose.

Let us consider some quite specific concerns under these three headings.

THE IMPORTANCE OF AGRICULTURE

In previous chapters we traced the way in which developed agricultural systems have tended to substitute support energy, chiefly in the form of machines and fuel, for man-power and animal power. Other inputs, such as fertiliser, involve a lot of energy in their manufacture and result in higher yields per unit of land and per man.

The net result of all this is to reduce greatly the number of people directly involved in agriculture: it is difficult to assess the effect on the total numbers involved, including those concerned with the manufacture and distribution of all the inputs to agriculture. It is even more difficult to calculate the numbers involved in food processing and distribution and the manufacture of the cans and vans employed in such activities, or to interpret such figures.

What is clear, however, is (a) that at one time, most people must have been closely concerned with food production and (b) that in recent years, in the industrialised countries, it has looked as though the production of food could be left as a chore for a minority of the population, whilst the rest got on with more important, more interesting or more progressive activities.

All this leaves out of account the people engaged in producing the food that is imported, or the people involved in transporting it and distributing it to the consumer.

Whether people really preferred to be engaged in non-food producing activities can hardly be disentangled from the relatively higher wages and other rewards associated with such activities. Certainly, it is curious how many people spend much time gardening and cultivating allotments and it could be that many people would be delighted to engage in food production, at least on a part-time basis, if this made economic sense.

At the present time, the world is divided between countries that are overproducing major food commodities, at considerable cost, and countries that are much less than self-sufficient in food production, many of them unable to afford the necessary imports.

In the EEC and USA, production of the main foodstuffs greatly exceeds demand and it has not proved possible to solve the problem by exports (since those in greatest need do not have the money), by food aid (because of failure of such aid to feed the hungry, because of the risks of corruption and of putting the local producer out of business when food is given away to his customers), or by increased home consumption.

In the EEC, the cost of agricultural support and of storage and disposal of surpluses has become a major difficulty, threatening the whole Common Agricultural Policy. All this reflects the original determination that agricultural production must never again be allowed to stagnate. It is quite remarkable how production increases, in virtually all countries, in response to real incentives. Unfortunately, it does not follow that the benefit goes to the farmers: much of it becomes capitalised in high land values which, in turn, place a high-cost burden on farming.

In such circumstances, countries with surpluses are unlikely to import foods from countries that are predominantly agricultural and need to export in order to fund development.

Countries that are short of everything find it difficult to provide incentives to their farmers: in fact, they are more likely to decrease incentives in order to keep down the price of food to the urban populations.

With such problems, it is difficult to appreciate that agriculture is usually by far the most important industry, even in big, industrialised countries. Few other industries (and these are usually rather arbitrary aggregations) contribute more to the gross national product or even employ more people.

TABLE 15.1
PROPORTION OF WORKING POPULATION
EMPLOYED IN AGRICULTURE 1986

Country	%
UK	2·3
USA	2·7
Australia	5·7
New Zealand	9·9
Cuba	21·0
Mexico	32·5
Egypt	42·5
India	67·8
Tanzania	82·9

Source: FAO (1987)

Even in the UK, which probably employs a smaller proportion (2·3%) of its economically active population directly in agriculture, is still a major employer and the total industry generates over one million jobs. Table 15.1 shows the proportion employed in agriculture in a range of different countries.

Agriculture is usually important for three different kinds of reason:

(1) The products it produces (and not only food), their volume, value, quality, consistency and safety;
(2) the employment generated, directly and indirectly;
(3) the role it has in managing the countryside. Most countrysides would look different—often less attractive and less accessible—if farming was not practised.

That is not to say that its effects are always wholly beneficial.

THE EFFECTS OF AGRICULTURE ON THE NON-AGRICULTURAL COMMUNITY

The most obvious effects, beyond such issues as the supply of food and raw materials, are on visual amenity, pollution and the shape and structure of the environment.

Visual amenity is influenced by the very fabric of farming, the shape of fields, the removal of hedges, the presence of tower silos, the crops grown and the absence of weed flowers, and even the colour of the animals and whether they are visible in the fields or kept indoors.

Pollution may take many forms: the smell of manure, the noise of grass-driers, nitrate in water courses and mud on roads, are all fairly obvious examples.

Other effects on the shape and structure of the environment include blocking of footpaths, landscaping of reservoirs and planting of shelter belts.

More subtle influences also occur, such as the interactions between farming methods and non-farmers' attitudes, well illustrated by the burning of straw and animal welfare. In the case of straw-burning, popular opinion may object to what seems to be unnecessary waste coupled with a fire hazard and it is obvious to many that it is happening. Objections to methods of keeping calves in dim light or poultry crowded into battery cages, however, do not depend on the conditions being visible at all, and quite different worries lie behind objections to the widespread use of pesticides, for example.

Some of these attitudes serve as constraints on agricultural methods but there are also positive attitudes encouraging development towards the provision of additional facilities. This is especially so in relation to recreation.

RECREATIONAL FACILITIES

Agricultural activities themselves are sometimes of interest to non-agriculturalists and visitors will pay to see such ordinary activities as milking. Zoo parks are an extreme form of this kind of catering for visitors and may be integrated with the conservation of wild life and plants and breeds of farm animals that are becoming rare.

Agricultural land is a major source of recreational facilities such as walking, shooting, hunting and fishing, and provides sites for camping and caravans.

The integration of small farm businesses and various forms of tourism is often of great benefit to all concerned and it might be better if the same integrated approach could be taken to many different forms of land use.

AGRICULTURE AND THE CITIZEN

It is clear that the interactions between agriculture and the food industry, its dominant role in land use and its major effect therefore on the environment of almost everyone, and its effect on the diet of the consumers, should make agriculture of concern to all citizens, whether they are directly engaged in it or not. This is as true today as when Socrates said 'No man qualifies as a statesman who is entirely ignorant of the problems of wheat'.

Yet this is not a general view and there are now many whose knowledge or understanding of agriculture is very small indeed. Of course, we are all ignorant of most industries, but agriculture does affect some of the most important parts of the lives of all citizens—food and environment.

Some time ago (1959), C. P. Snow (1964) described, in his Rede Lectures, what he called the 'two cultures' resulting from early specialisation in either arts or science subjects. He argued that a complete education required both and suggested two criteria that could be used to assess the possession of such an education.

Snow postulated that no-one could claim to be educated unless, as a minimum:

(1)   He could describe the Second Law of Thermodynamics.†
(2)   He had read a work of Shakespeare.

Whatever one may think of the proposition in general or these two criteria in particular, it is tempting to add as a third requirement that no-one can be regarded as educated if he thinks that milk comes from a bottle. This, of course, is a rather trite way of expressing it and it is very revealing to try and express positive criteria that really fill the need.

Two important criteria suggest themselves, in the following terms.

First, that one understands that photosynthesis *fixes* solar energy.

Secondly, that one has heard of nitrogen fixation by rhizobia in the roots of legumes.

However, both of these are essentially biological, rather than agricultural, and the most useful, characteristically agricultural, criterion may be an understanding of the role of the ruminant in feeding people. Having regard to the rather sweeping nature of the proposal that without such an understanding no-one is agriculturally educated and cannot claim to be a complete citizen, it is worth considering carefully whether it is reasonable and what other criteria might serve. In considering how the ordinary citizen's education could or should be influenced by agriculture, it should not be supposed that the agriculturalist's education is necessarily satisfactory.

As mentioned in Chapter 2, there is a great need for those who can view agricultural systems as a whole, without being bounded by their disciplinary training. Yet most agricultural education is organised in this way and most graduates tend to be specialists in crop production, plant physiology, veterinary medicine, soil science, economics, animal production, parasitology or one of a host of other specialisms.

Not that specialists are not needed, but their ability to apply what they know would be greatly improved if they also had some capacity to use a systems approach, both to the definition and the solutions of the problems they study.

Some changes are being made. Universities such as Reading have long included some systems thinking in the degree in agriculture. The Open University in the UK has applied it to whole courses and Hawkesbury

† This is variously stated as, for example: 'The entropy or degree of randomness of the universe tends to increase' and 'No energy conversion system is ever completely efficient'.

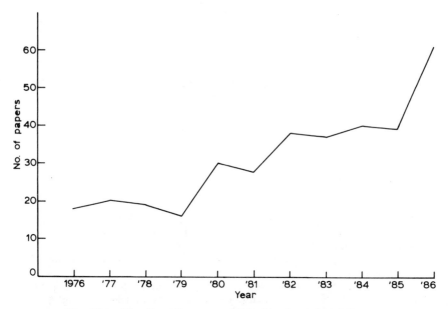

FIG. 15.1.   Number of papers published in *Agricultural Systems*.

College, NSW, Australia, has based a whole degree on it. Postgraduate research on a systems basis is now carried out at many universities: taught postgraduate courses are now provided, for example, in Thailand. Thus the systems approach is slowly spreading, entirely based on the simple recognition that farms and agricultural enterprises are systems and best understood by thinking of them in this way.

The progress in research and development is illustrated by the steady increase in the number of papers published by the international journal 'Agricultural Systems' since it was launched in 1976. These papers (see Fig. 15.1) have come from nearly 40 countries and represent a steady increase in agricultural research based primarily on a systems approach.

Bibliography

AUSTIN, R. B. (1975). Economy in the use of manufactured fertilisers. Use of Energy in Agriculture. Annual Conference of The Institute of Agricultural Engineering.

BATHER, D. M. (1975). Personal communication.

BØRGSTROM, G. (1973). *World Food Resources*. International Textbook Co., Aylesbury.

CARRUTHERS, S. P. and JONES, M. R. (1983). Biofuel production strategies for UK agriculture. CAS Paper 13. Centre for Agricultural Strategy, Reading.

COOKE, G. W. (1975). *Fertilising for Maximum Yield* (2nd Ed.). Crosby Lockwood Staples, London.

COOKE, G. W. (1983). Constraints imposed by the soil on crop production. Review Article. Letcombe Lab. Ann. Rep. 1982. 84–102.

COOPER, J. P. (1976). Photosynthesis as an industrial source. *Proc. Conf. Birmingham Univ.* 1975. *Energy and the Environment.* 99–118.

DAVIDSON, H. R. and COEY, W. E. (1966). *The Production and Marketing of Pigs* (3rd Ed.). Longman, London.

DEBACH, P. (1964). Successes, trends and future possibilities. In: *Biological Control of Insect Pests and Weeds* (DeBach, P. (Ed.)). Chapman and Hall, London, 673–713.

DEBACH, P. (1974). *Biological Control by Natural Enemies.* Cambridge University Press.

DILLON, J. L. (1977). *The Analysis of Response in Crop and Livestock Production.* (2nd Ed.). Pergamin Press.

DUCKHAM, A. N. and MASEFIELD, G. B. (1970). *Farming Systems of the World.* Chatto and Windus, London.

DUCKHAM, A. N. (1976). Environmental constraints. In: *Food Production and Consumption* (Duckham, A. N., Jones, J. G. W. and Roberts, E. H. (Eds.)). North-Holland, Amsterdam.

EDWARDS, C. A. and HEATH, G. W. (1964). *The Principles of Agricultural Entomology.* Chapman and Hall, London.

FAO (1977). *Production Yearbook, 1976.* Vol. 30. FAO, Rome.

FAO (1985a). *Production Yearbook, 1984.* Vol. 38. FAO, Rome.

FAO (1985b). *State of Food and Agriculture, 1984.* FAO, Rome.

FAO (1987). *Production Yearbook, 1986.* Vol. 40. FAO, Rome.

178

FARMER, D. S. and KING, J. R. (1971). *Biology*. Vol. 1. Academic Press, London and New York.

Farmer's Weekly (1976). Explosives, buttons and aphrodisiacs. Sept. 17 (IX)–(XVI).

FERNANDES, E. C. M. and NAIR, P. K. R. (1986). An evaluation of the structure and function of tropical home gardens. *Agric. Syst.*, **21**, 279–310.

GRIGG, D. B. (1974). *The Agricultural Systems of the World. An Evolutionary Approach*. Cambridge University Press.

GUGGENHEIM, K., WEISS, Y. and FOSTICK, M. (1962). Composition and nutritive value of diets consumed by strict vegetarians. *Brit. J. Nutr.*, **16**, 467–74.

HAFEZ, E. S. E. (1968). *Adaptation of Domestic Animals*. Lea and Febiger, Philadelphia.

HARRISON, G. A., WEINER, J. S., TANNER, J. M. and BARNICOT, N. A. (1964). *Human Biology*. Oxford University Press.

HEYDEMAN, M. T. (1973). Aspects of protein production by unicellular organisms. In: *The Biological Efficiency of Protein Production* (Jones, J. G. W. (Ed.)). Cambridge University Press, 303–22.

HOMB, T. and JOSHI, D. C. (1973). The biological efficiency of protein production by stall-fed ruminants. In: *The Biological Efficiency of Protein Production* (Jones, J. G. W. (Ed.)). Cambridge University Press, 237–62.

JANICK, J., SCHERY, R. W., WOODS, F. W. and RUTTAN, V. W. (1969). *Plant Science: An Introduction to World Crops*. W. H. Freeman and Co., San Francisco.

LARGE, R. V. (1973). Factors affecting the efficiency of protein production by populations of animals. In: *The Biological Efficiency of Protein Production* (Jones, J. G. W. (Ed.)). Cambridge University Press, 183–200.

MATLOCK, W. G. and COCKRUM, E. L. (1974). *A Framework for Evaluating Long-term Strategies for the Development of the Sahel–Sudan Region*. Vol. 2. *A Framework for Agricultural Development Planning*. Camb., Mass. Center for Policy Alternatives, M.I.T.

MELLANBY, K. (1975). *Can Britain Feed Itself?* Merlin Press, London.

MORRIS, D. (1965). *The Mammals*. Hodder and Stoughton, London.

NGAMBEKI, D. S. (1985). Economic evaluation of alley cropping Leucaena with Maize–Maize and Maize–Cowpea in Southern Nigeria. *Agric. Syst.*, **17**, 243–58.

NIX, J. (1976). *Farm Management Pocketbook*. (7th Ed.). Wye College, Univ. of London.

NORMAN, L. and COOTE, R. B. (1971). *The Farm Business*. Longman, London.

NYE, P. H. and GREENLAND, D. J. (1960). *The Soil Under Shifting Cultivation*. C. A. B. Harpenden. Tech. Comm. No. 51.

NEWTON, J. E. and BETTS, J. F. (1966). Factors affecting litter size in the Scottish half-bred ewe, I, *J. Reprod. Fert.*, **12**, 167–75.

NEWTON, J. E. and BETTS, J. F. (1968). Factors affecting litter size in the Scottish half-bred ewe, II, *J. Reprod. Fert.*, **17**, 485–93.

OKIGBO, B. N. (1977). Legumes in farming systems of the humid tropics. In: *Biological Nitrogen Fixation in Farming Systems of the Tropics*. (Ayanaba, A. and Dart, P. J. (Eds)). John Wiley and Sons, Chichester.

PIRIE, N. W. (1971). *Leaf Protein: Its Agronomy, Preparation, Quality and Use*. IBP Handbook No. 20. Blackwell Scientific Publications, Oxford and Edinburgh.

PIRIE, N. W. (1987). *Leaf Protein and Its By-products in Human and Animal Nutrition.* Cambridge University Press.

RAPP, A., LE HOUEROU, H. N. and LINDHOLM, B. (1976). *Can desert encroachment be stopped?* Ecological Bulletin No. 24. Report published by UNEP and SIES.

RUTHENBERG, H. (1971). *Farming Systems in the Tropics.* Oxford University Press.

SIMMONDS, N. W. (1985). Farming systems research. A review. World Bank Technical Paper No. 43. World Bank, Washington.

SNOW, C. P. (1964). *The Two Cultures: And a Second Look.* Cambridge University Press.

SPEDDING, C. R. W. (1970). *Sheep Production and Grazing Management.* (2nd ed.). Bailliere, Tindall and Cassell, London.

SPEDDING, C. R. W. (1971). *Grassland Ecology,* Oxford University Press.

SPEDDING, C. R. W. (1976a). In: *Energy and the environment. Proc. Conf. at Birmingham Univ.* in 1975, 337–55.

SPEDDING, C. R. W. (1976b). An assessment of the future developments of green crop fractionation in the UK. *Proc. BGS/BSAP Symp., Harrogate.*

SPEDDING, C. R. W. and HOXEY, A. M. (1975). The potential for conventional meat animals. In: 'Meat'. Univ. of Nottingham Easter School, 1974.

SPEDDING, C. R. W. and WALSINGHAM, J. M. (1975). The production and use of energy in agriculture, *J. Agric. Econ.* **XXVII**, 19–30.

SPEDDING, C. R. W. (1984). New horizons in animal production. Chapter 28. In: *Development of Animal Production Systems* (B. Nestel (Ed)). Vol. A2. World Animal Science, Elsevier.

SPEDDING, C. R. W. (1985). What will the public demand? In: *The Future Demands Change.* Proc. 39th Oxford Farming Conf. 1985.

TATCHELL, J. A. (1976). Crops and fertilisers; overall energy budget. In: *Conservation of Resources,* special publication No. 27. The Chemical Society.

USDA (1980). Report and Recommendations on Organic Farming. USDA, Washington.

USSAC (1967). The world food problem. Report to the President's Advisory Committee by the United States Science Advisory Committee. US Government Printing Office, Washington DC.

VAN DEN BOSCH, R. (1971). Biological control of insects. *Ann. Rev. Ecol. and Syst.,* **3,** 45–66.

VAN EMDEN, H. F. (1974). *Pest Control and Its Ecology.* Inst. Biol. Studies No. 50. Edward Arnold.

WALKER, T. (1973). Protein production by unicellular organisms from hydrocarbon substrates. In: *Biological Efficiency of Protein Production* (Jones, J. G. W. (Ed.)). Cambridge University Press, 323–38.

WEATHERLEY, A. H. (1972). *Growth Ecology of Fish Population.* Academic Press.

WEBSTER, C. C. and WILSON, P. N. (1966). *Agriculture in the Tropics.* (2nd Ed.). Longman, London.

WHITE, D. J. (1975). The use of energy in agriculture. *Farm Management,* **2**(12), 637–50.

WILSON, P. N. (1973). The biological efficiency of protein production by animal production enterprises. In: *The Biological Efficiency of Protein Production* (Jones, J. G. W. (Ed.)). Cambridge University Press, 201–10.

WOKES, F. (1956). Anaemia and vitamin B_{12} dietary deficiency. *Proc. Nutr. Soc.*, **15**, 134–41.

WORGAN, J. T. (1973). Protein production by micro-organisms from carbohydrate substrates. In: *Biological Efficiency of Protein Production* (Jones, J. G. W. (Ed.)). Cambridge University Press, 339–62.

WORGAN, J. T. (1976). The efficiency of technological systems of food production. In: *Food Production and Consumption* (Duckham, A. N., Jones, J. G. W. and Roberts, E. H. (Eds)). North Holland, Amsterdam, 215–30.

ZUENER, F. E. (1963). *A History of Domesticated Animals*. Hutchinson, London.

Further Reading

Chapter 1.   1. *The Biology of Agricultural Systems.*
            C. R. W. Spedding, 1975. Academic Press.
         2. *An Introduction to Farming Systems.*
            Michael Haines, 1982. Longman.
         3. *Managing Agricultural Systems.*
            G. E. Dalton, 1982. Applied Science Publishers.

Chapter 2.   1. *Systems Analysis in Agricultural Management.*
            Eds J. B. Dent and J. R. Anderson, 1971. John Wiley & Sons
            Australasia Pty Ltd.
         2. *The Study of Agricultural Systems.*
            Ed. G. E. Dalton, 1975. Applied Science Publishers.
         3. *Simulation of Ecological Processes.*
            C. T. de Wit and J. Goudriaan, 1974. Pudoc, Wageningen.
         4. *Computer Modelling in Agriculture.*
            N. R. Brockington, 1979. Clarendon Press, Oxford.
         5. *Higher-yielding Human Systems for Agriculture.*
            W. F. Whyte and D. Boynton, 1983. Cornell University Press.
         6. Spedding, C. R. W. and Brockington, N. R., 1976.
            Experimentation in agricultural systems. *Agric. Sys.* **1**, 47–56.

Chapter 3.   1. *Farming Systems of the World.*
            A. N. Duckham and G. B. Masefield, 1970. Chatto & Windus.
         2. *The Agricultural Systems of the World.*
            D. B. Grigg, 1974. Cambridge University Press.

Chapter 4.   1. *Grassland Ecology.*
            C. R. W. Spedding, 1970. Oxford University Press.
         2. *The Biological Efficiency of Protein Production.*
            Ed. J. G. W. Jones, 1973. Cambridge University Press.
         3. *Biological Nitrogen Fixation in Farming Systems in the Tropics.*
            Eds A. Ayanaba and P. J. Dart, 1977. John Wiley.
         4. *Biological Efficiency in Agriculture.*
            C. R. W. Spedding, J. M. Walsingham and A. M. Hoxey, 1981.
            Academic Press.

5. *Agricultural Ecosystems: Unifying Concepts.*
   Eds R. Lowrance, B. R. Stinner and G. J. House, 1984. John Wiley.

Chapter 5. 1. *Agricultural Resources.*
   A. M. Edwards and A. Rogers, 1974. Faber.

   2. *Agricultural Economics.*
   C. Ritson, 1977. Crosby Lockwood Staples.

Chapter 6. 1. *Criticism and the Growth of Knowledge.*
   Eds Imre Lakatos and Alan Musgrave, 1970. Cambridge University Press.

Chapter 7. As for Chapter 3.

Chapter 8. 1. *Farming Systems in the Tropics.*
   Hans Ruthenberg. (2nd Ed.) 1977. Oxford University Press.

   2. *Traditional African Farming Systems in Eastern Nigeria.*
   J. Lagemann, 1977. Weltforum Verlag Munchen.

   3. *Annual Cropping Systems in the Tropics.*
   M. J. T. Norman, 1979. University of Florida.

Chapter 9. 1. *Farming Systems in the Tropics.*
   Hans Ruthenberg. (2nd Ed.) 1977. Oxford University Press.

Chapter 10. 1. *Farming Systems of the World.*
   A. N. Duckham and G. B. Masefield, 1970. Chatto & Windus.

   2. *The Farm Business.*
   L. Norman and R. B. Coote, 1971. Longman.

   3. *The Ecology of Agricultural Systems.*
   T. P. Bayliss-Smith, 1982. Cambridge University Press.

Chapter 11. 1. *The Biology of Agricultural Systems.*
   C. R. W. Spedding, 1975. Academic Press.

   2. *Principles of Field Crop Production* (3rd Ed.)
   J. H. Martin, W. H. Leonard and D. L. Stamp, 1976. Macmillans.

Chapter 12. 1. *The Biology of Agricultural Systems.*
   C. R. W. Spedding, 1975. Academic Press.

   2. *Animals for Man.*
   J. C. Bowman, 1977. Edward Arnold.

   3. *Animal Agriculture.*
   H. H. Cole and M. Ronning, 1974. W. H. Freeman & Co.

Chapter 13. 1. *Single Cell Protein.*
   Eds R. I. Mateles and S. R. Tannenbaum, 1968. M.I.T. Press.

Chapter 14. 1. *Food Production and Consumption.*
   Eds A. N. Duckham, J. G. W. Jones and E. H. Roberts, 1976. North Holland & American Elsevier.

   2. *Biological Efficiency in Agriculture.*
   C. R. W. Spedding, J. M. Walsingham and A. M. Hoxey, 1981. Academic Press.

   3. Spedding, C. R. W., 1984. Energy use in the food chain. *Span* **27**(3), 116–118.

Chapter 15. 1. *How the Other Half Dies.*
   Susan George, 1977. Penguin Books.

2. *Man and His Environment: Food.*
   L. R. Brown and G. N. Finsterbusch, 1972. Harper & Row.
3. *World Food Resources.*
   G. Børgstrom, 1973. Intertext Books.
4. *Pesticides and Pollution.*
   K. Mellanby, 1967. Fontana.
5. *World Food Resources.*
   Michael Allaby, 1977. Applied Science Publishers.

Index